BestMasters

Springer awards "BestMasters" to the best application-oriented master's theses, which were completed at renowned chairs of economic sciences in Germany, Austria, and Switzerland in 2013.

The works received highest marks and were recommended for publication by supervisors. As a rule, they show a high degree of application orientation and deal with current issues from different fields of economics.

The series addresses practitioners as well as scientists and offers guidance for early stage researchers.

Enrico Marcantoni

Collateralized Debt Obligations

A Moment Matching Pricing Technique based on Copula Functions

Springer Gabler

Enrico Marcantoni
Vienna, Austria

Masterthesis, University of Applied Sciences (bfi) Vienna, Austria / University Bologna, Italy

ISBN 978-3-658-04845-7   ISBN 978-3-658-04846-4 (eBook)
DOI 10.1007/978-3-658-04846-4

The Deutsche Nationalbibliothek lists this publication in the Deutsche Nationalbibliografie; detailed bibliographic data are available in the Internet at http://dnb.d-nb.de.

Library of Congress Control Number: 2013958235

Springer Gabler

Printed on acid-free paper

Springer Gabler is a brand of Springer DE.
Springer DE is part of Springer Science+Business Media.
www.springer-gabler.de

Foreword

The thesis deeply analyzes the existing approaches to price CDOs concerning the modeling of the individual default probability, the loss given default and particularly the default correlation between obligors. The most common approaches to recover the default correlation are represented by the reduced approach, where the dependencies of obligors are written in term of default intensities, and the structured approach based on the seminal paper by Merton (1974). Here the default correlations are expressed through the dependencies of the assets into the firm's portfolio on some common factors which allow to consider the granularity and the sectorial concentration and contagion risk. Obviously as the number of common factors increases, the valuation becomes more precise but the complexity of the model considerably increases. For this reason the reference paper of this thesis is Castagna et al. (2012) where, in a structured multi-factor model, they provide a closed-form for pricing CDO based on a Moment Matching technique.

However the original work of Castagna et al (2012) models the default correlation through a multivariate Gaussian copula function which does not allow to consider more general asymmetric dependency structures with tail dependencies; then the model has been rewritten here in an Archimedean framework such to propose a more general representation of the dependencies. Moreover the contribution of this thesis is not only theoretical because here for the first time the Moment Matching technique has been implemented to price a CDO. The data refer to a CDX composed by 125 names whose CDSs' quotes were collected on July 3rd 2007. As the default probabilities have been bootstrapped from the CDSs' prices with maturity 10 years, based on these marginal probabilities, a clustering approach allows to reduce the complexity in the numerical computation of the moments of the original distribution of the portfolio loss. The idea of the Moment Matching technique is to calibrate the moments of the original distribution on the moments of an approximated one, assumed to be the well known Vasicek's limiting model here, such to recover the parameters of the approximated distributions and then the tranches' price. It's very interesting to point out that the moment expansion that is very useful for pricing purposes, allows also to identify the potential risk sources entering in the pricing process and to construct an hedging portfolio to cover them.

Bologna, October 17, 2013

Prof. Silvia Romagnoli

Contents

List of Figures

List of Tables

List of Abbreviations

ABS	Asset Backed Securities
ANN	Artificial Neural Network
ASRF	Asymptotic Single Risk Factor
Bp	Basis point
CBO	Collateralized Bond Obligation
CDF	Cumulative Distribution Function
CDO	Collateralized Debt Obligation
CDS	Credit Derivative Swap
CDX	Credit Default Swap Index
CLN	Credit Link Note
CLO	Collateralized Loan Obligation
EAD	Exposure at Default
EC	Economic Capital
ETL	Expected Tranche Loss
GCLHP	Gaussian Copula Large Homogeneous Portfolio
LGD	Loss Given Default
LLN	Law of Large Numbers
MLE	Maximum Likelihood Estimate
MM	Moment Matching
PD	Probability of Default
RWA	Risk weight asset
SOM	Self Organized Map
SPV	Special purpose vehicle
VaR	Value at Risk

Abstract

The thesis focuses on a method to price CDO tranches. The original method is developed by Castagna *et al.* in 2012.

The thesis numerically implements the methodology to price a CDX. In addition the original model is then extended in terms of Clayton copula function. In particular, an extension of the model in terms of Clayton copula is provided by proposition 6.1, 6.2 and 6.3. The moment expansion, presented in the original model, is here used as method to highlight and manage sources of risk.

Finally, the Clayton extension will be implemented to the same set of data and the different tranche prices obtained will be compared. It follows that the different characteristics of the two copula function are evident only in the senior and super senior tranche.

1   Introduction

1.1   Selection and relevance topics

The financial crisis which started in 2007 has shown a comprehensive undervaluation by the Financial Institutions, of the risk involved in credit derivatives, such as collateralized debt obligations (CDOs).

The complexity of CDOs, combined with inadequate tools for modeling the risk, solicited the formation of a more robust approach to measure and price them.

In pricing CDOs, the main problem is represented by the fair evaluation of the tranches premium. Similarly any other instrument, pricing the tranches premium, means fairly compensating the investor for the eventual expected losses. The latter is strictly linked to the individual default probability (PD), the loss given default (LGD) and default correlation between obligors of the reference portfolio. Modeling the latter represents the main focus in pricing CDOs and it is performed following two main approaches: the reduced and the structured approach.

The first relies on market price of defaultable firms and the default correlation is obtained building the dependencies of obligors in terms of default intensities. The second, following the work of Merton (1974), takes in consideration the main firms fundamentals (equity and debt), and the default correlation is expressed through the exposure of the firm assets dynamics on one or more common factors. The latter approach is also the one adopted by Basel II aiming to compute the Credit VaR. In particular, this approach is called the Asymptotic Single Risk Factor (ASRF) and it is based on the work of Vasicek (1991). The ASRF does not take into consideration the sector and contagion effect, aiming to give an easy and fast implementation methodology to compute VaR..

In this framework the recent literature extended the ASRF model, by introducing such risk sources. Bonollo *et al*. (2009) extended the basic ASRF hypothesis by considering granularity risk, a sectorial concentration and a contagion risk. This work has been generalized by Castagna *et al*. (2009) to a multi-scenario setting. Both works derive analytical approximations working very efficiently for the computation of the Credit VaR, which is computed at a high level of confidence. However, this efficiency is lost working at lower level of confidence, as the ones required in the pricing of a CDO.

In this framework, Castagna *et al.* (2012) starting from their work of 2009, developed a specific method, based on a Moment Matching techniques, for pricing CDO.

The importance of the model relies at the same time on two aspects. Firstly, sharing the same framework of the previous work of 2009, the model is consistent with the Credit VaR computation. This enables financial institutions to have a unified approach to both evaluate the Credit VaR and the risk of structured products they issue. Secondly, the model is based on a more realistic hypothesis in contrast to the ASRF, but at the same time provides a closed-form which makes the implementation easy and quick, opposed to the more complex methods.

In the thesis the work of Castagna *et al.* (2012) is for the first time implemented in the pricing of a CDO. Furthermore, in the thesis an extension of the model is provided. In the extension, the original model has been rewritten in terms of Archimedean copula. This extension has been implemented as well, and the results have then been compared.

1.2   Formulating the research questions

The thesis focuses on the presentation, implementation and extensions of the work of A.Castagna, F.Mercurio, P.Mosconi, "Analytical Pricing CDOs in a Multi-factor Setting by a Moment Matching Approach". Four research questions represent the focus of the thesis.

The original work of Castagna *et al.* has not been applied to the real data and it is here implemented for the first time. This leads to the first research question. Is it possible to obtain the tranches prices of a CDO by implementing the model?

The second research question is strictly linked to the moment expansions used in the model. Is it possible to determine potential risk sources and implement the method to cover them?

The third and the fourth research questions refer to possible extensions of the model. In the original models, the default correlation is modeled through multivariate Gaussian copula. The third research question follows. Is it possible to generalize the default correlation in terms of any copula functions and in particular is it possible to rewrite the model in a Archimedean framework using the Clayton copula?

2

The fourth, and last research question, consequently follows. Is it possible to implement the Archimedean model and what are the differences in term of tranches price obtained?

1.3   State of the literature

The default correlation and its modeling through the use of copula function has received notable interest during the last decades both by the industry side, given the misjudging of risk highlighted by the financial crisis, as well by the academia side, given the open issue these still represent.

As already reported the default correlation is modeled by two main approaches: the reduced and the structured approach. The latter is mainly due by Merton (1974). This model represents the starting point of all the subsequent literature, focusing on the default correlation. In particular, the default correlation is expressed as the exposure of asset dynamics on one or more common factors. This approach is also the same applied by the international regulation in Basel II for the Credit VaR computation. It relies on the ASRF based on the previous work of Vasicek in 1991. The success of this approach is due to its closed-form, which makes the implementation quick and fast. However, the model is based on strong hypothesis, such the homogeneity, which make the model inconsistent with the reality.

In this context several authors tried to extended the model, contributing to create a huge and important literature.

The extensions of the model are directed in two directions: one aiming to make the general hypothesis of the model stronger by eliminating the homogeneity assumption and the other aiming to modify the obligors dependencies, but holding the homogeneity assumption. Concerning the first point, Bonollo *et al.* (2009) and Castagna *et al.* (2009) and (2012), both derived efficient analytical Credit Var approximations remote from the homogeneous portfolio assumption.

Relating to the second point, Schonbucher (2003) extends Vasicek's work to Archimedean copulas. Here, the homogeneity assumption holds, but the obligors dependencies is based on Archimedean copulas.

1.4 Methodology

To begin with, the main theory aspect will be presented in the first chapters. In particular these represent the necessary requirements for the focus of the thesis.

Once these notions will be give, the thesis will concentrate in a practical implementation aiming to price a CDX. The implementation is based on real data.

Once the historical data are collected, through Microsoft Excel Spreadsheets created by the author, the prices of the CDX tranches are obtained.

1.5 The structure of the thesis

After this introductory chapter the structure of the thesis will be as follows.

In the next chapter the CDO main characteristics will be explained. Once the general structure of such contracts will be reported, a description of the main subjects involved in such contracts are then presented. A classification of the different kind of CDOs is then reported and in particular, being the application based on a CDX, such indices will in detail analyzed. The chapters concludes with the general CDO pricing formulas.

Once given a description of CDO characteristics, the main models based on correlated defaults will be presented in chapters 3. This chapter offers a detailed overview of the past credit risk modeling literature. It starts from the easiest model, such as the Bernoulli Model arriving to the one and multi-common factors models. The chapter presents the loss dependence in terms of copula functions.

Chapter 4 gives the main notions of copula functions. To begin with, the definition of copula and the fundamental theorem of Sklar will be given. The presentation of main copula families and their main characteristics then follow. The concept of tail dependence is then explained.

After these theory chapters, which represent the necessary notions to understand the model framework and its extensions, the main targets of the thesis will be reached in the subsequent chapters.

In chapter 5 the original model of Castagna *et al.* is presented. In particular, the only part reported is the Moment Matching procedure, which is a technique to obtain tranches price of a CDO.

Once the method has been presented, the contribution of the thesis to existing literature will be given in the following chapters.

Two extensions of the model are derived in chapter 6, Firstly, the model is rewritten in terms of Clayton copula. The proposition 6.1 , 6.2 and 6.3, derived in the thesis, allows a rewriting of the model in terms of a Clayton dependency. Secondly, the moment expansions, derived in the original model, is here used to present a method to managing risk sources.

Once these extensions are derived, the numerical implementation follows in the last chapter. Chapter 7 is divided in three section. In the first section, the original method of Castagna *et al.* is implemented. This represents a first contribution to the original work, given that such method has not been applied to real data. In the second section, the new method derived in the previous chapter, is applied to real data as well. In the last section, the results obtained using the two method, and so the two dependency structures, are compared.

2 CDO: general characteristics

2.1 Introduction and definition

Securitization has started to play a huge role in the market of the structured products since the beginning of the 1990s, reaching significant levels in the last decade. The terms "securitization" refers to the transfer of an asset pool into tradable securities.

In the securitization environment, CDOs, as their market volume demonstrate, represent one of the most popular instruments.

CDOs are a class of asset backed securities (ABS) which are securities backed by a pool of assets. Once the basket of assets is securitized it is available to be traded and the aim of a CDO is to allow the *buy side* to reach risk-return profile, otherwise not available under regulation, and the *sell side* to discard their risk exposures[1].

In a CDO deal, assets exposed to credit risk exposure are pooled and sold to a juridical subject called the *special purpose vehicle* (SPV). The SPV invests in the diversified pool of assets financing the purchase via the issue of financial instruments called notes, which are sold to the market in different tranches. Tranches are classified according to their grade of seniority and to their different risk-return trade off. The seniority refers to the priority in both repayment of the principal and payment of the interest. Usually, the "senior tranche" has a rating between AAA and A and it is the one with the highest priority. The "mezzanina tranche", with a rating between BBB and B and the "equity tranche", follows. The latter, sometimes also called equity piece, is the one with the highest risk profile, being the first tranche absorbing all the possible losses, and for this reason is usually unrated. According to this procedure the losses on the asset side are transformed in losses on the tranches, and the tranches holders entirely absorb them. This process is also called "waterfall".

2.2 The tranches role

CDO is an operation of structured finance which recurs on the tranches use. Suppose the *sell side* holds a portfolio, constituted by a set of defaultable

---

[1] Cherubini et al. (2007) p. 203.

instruments of different firms and its aim is discarding the risk exposure coming from that pool. The *sell side*, assuming it cannot sell the entire portfolio, could cut off the risk exposure. One way is represented by buying a credit default swaps (CDS) on each name the instruments refers to. Another possibility, the one this work is based on, is represented by a CDO. That is, tranching the portfolio and selling the credit risk incorporated in the tranches.

Tranches are the main characteristic on which structured finance operations are based. Each tranches is characterized by a lower limit called attachment point (L) and an upper limit called detachment point (U). These express the percentage of the total portfolio loss covered by the tranches. Usually the equity tranche is characterized by $L = 0$ and $U > 0$. The detachment point for each tranche overlaps with the attachment point of the subsequent, in terms of seniority degree, tranche. An holders of a tranche with attachment points $[L, U]$ is responsible for the asset pool losses exceeding L up to U. In this way the holder of that tranche will not suffer any loss when the total portfolio losses are lower than L and he will not be liable for the part of losses greater than U.

The tranches holders, bearing the credit risk included on the backed asset, have to be compensated via a periodic premium along the life of the CDO. Obviously, the lower the tranche seniority degree, the higher the tranche premium. That is, the highest premium is due to the equity tranche, which in several CDO contracts also receive a further initial upfront.

2.3 Classification of CDOs

Given the continuous products innovation and the slight differences existing between them, it is common to classify a CDO according to:

- the main aim of the operation in terms of the economic purpose the CDO would like to reach. It is possible to distinguish between balance sheet CDO, where the originator want to transfer credit risk, and arbitrage CDO, issued to profit from the difference between the market prices of the asset pool and the price of the securitized products.

- the management of the collateral as a way to manage notes 'cash flow and interests. It is possible to distinguish between cash flow CDO and market value CDO.

8

- the structure of the operation. It is possible to distinguish between cash CDO and synthetic CDO.

Finally it is also possible classify the CDOs according to the main characteristics of the asset constituting the pool. According to this criteria it is possible to distinguish between:

- collateralized bonds obligations (CBO), where the collateral pool are represented by credit risky bonds issued by both public or private firms. The main aim of CBO has to be found in arbitrage spread opportunities;
- collateralized loan obligation (CLO), where the collateral pool is represented by banks loans. Banks are mainly motivated by regulatory arbitrage altogether with economic risk transfer motivations.

2.4 Reasons for the utilization of CDOs

In the research of the reasons for the utilization of securitized products is fundamental to distinguish between the reasons of the financial institutions and the reasons of the general investors.

The most important reasons driving the financial institution to use securitized products have to be found in the risk management and in the diversification.

Supposing a bank's loan portfolio has a notable concentration in a certain industry or region, then the bank could reduce such concentration by securitizing part of the portfolio or by investing in securitized products concentrated on opposite region or industry.

Another important reason for the financial institutions lies in the regulatory capital relief.

In particular, taking their commercial loans as asset to securitize the main bank's aim is shrinking the balance sheet or reducing the capital requirements.

According to Basel II model, the regulatory capital requirement a bank has to dispose is 8% of the Risk Weight Assets (RWA) of the reference. After the tranching of the pool of loans, the new bank regulatory requirement is providing the capital corresponding to the piece retained. That is, passing the 8% of the RWA to the amount of equity piece retained.

However, whilst by one side this allows banks to benefit in terms of liquidity and risk transfer, by the other, there are implied huge costs of securitization that have to be take into consideration in the final decision

However, from a financial institution's point of view, securitization presents also drawbacks. To begin with, the benefit coming from regulatory capital relief might be limited by the holding of a large part of the equity tranches, which implies an incomplete transfer of credit risk exposure. Furthermore, securitization might be expensive considering legal costs, technical costs and rating agencies costs.

From a general investor point of view, securitization allow investor to reach risk-return adapt to his profile. For example, CDO tranches with the same ratings of other credit derivatives or bonds, have higher return. Furthermore, the credit risk exposure of structured products allow investor to achieve a credit risk exposure otherwise unreachable in the market. For this reason, some investors as institutional investor are not allowed to invest in such typologies of products.

2.5   Typical cash flow CDO structure

In this section, a typical cash flow CDO is explained, as illustrated in **Figure 1**. Although several classifications have been reported, only the cash and synthetic CDO are analyzed in detail below.

At the origin of a CDO operation there is a pool of asset, which could be remotely as well as only recently purchased by the bank, just to be inserted in the CDO pool. The originator is usually the holder of the assets, which sells it to the SPV.

The latter, as the name suggests, is a company set-up especially for the notes issue and the assets purchasing. The main characteristic of the SPV is the bankruptcy remoteness, reached via a strict legal separation between SPV and the originator, aiming to avoid a default of the SPV on its obligations due to bankruptcy of the originator.

The SPV, being a company expressly set-up, has no money for funding the purchasing of the assets, which will be purchased once the vehicle issued notes. That is, the total notional of the issued securities cover the principal of the pool. The interest and the principal due to notes investor will be covered with the interest and principal of the assets of the pool. Given that the investor is subject to the asset cash flow, investor purchasing notes absorb the risk of the pool. For this reason tranching notes in different classes, according to different risk profile, is comprehensible.

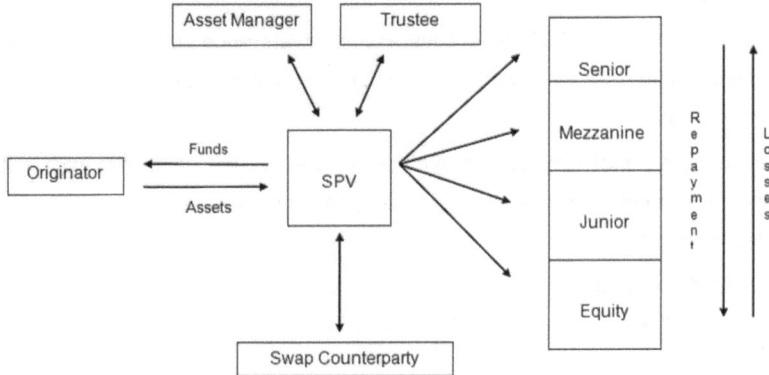

Figure 1: Typical cash flow CDO structure [2]

Alongside these two main figures, other subjects play important roles.

The *sponsor* is a subject interested in the realization of the operation. Usually the sponsor is the subject underwriting the equity piece. In balance sheet CDO the sponsor is always a bank aiming for asset restructuring while in the arbitrage CDO the sponsor could be either a bank or an intermediary, whose aim is earning by the commission fee.

The *arranger*, usually an investment bank, is the subject responsible for the tranches placement to investors. Its earning is represented by commission fee.

The *asset manager* is the subject responsible for collateral managing. In arbitrage CDO, in contrast of balance sheet CDO, this role is certainly more important, given that the collateral can be actively managed.

The *trustee* is the subject responsible to collect, on behalf of SPV, the cash flow of the collateral in order to pay the notes' interest and commission fees.

---

[2] Bluhm, Overbeck and Wagner (2003) p.287.

2.6 Synthetic CDO

Synthetic CDO appeared in the market at the end of the 1990s and, in the last years, they have become very popular.

Differently from the cash CDO, where the assets of reference portfolio are sold to the SPV which purchase the ownership rights, in the synthetic CDO only the credit risk is transferred.

The reduction of legal issues, and their associated costs, altogether with the flexibility of the structure for arbitrage and hedging aims, made synthetic CDO very requested instruments.

Synthetic CDO, according to the funding methodology used, can be divided into *fully funded synthetic CDO, unfunded synthetic CDO* and *partially funded synthetic CDO.*

In a fully funded synthetic CDO, as in **Figure 2**, the originator, that is the protection buyer transfers to the SPV, that is the protection seller, the credit risk of a reference portfolio through a CDS. As a CDS contract implies, the SPV receives from the originator a premium for the protection he has to pay if credit event occurs. The protection seller issues notes for a value par to the reference portfolio, tranching them in different risk classes. Investors, once notes are underwritten, themselves become protection sellers. The amount collected by the issuing is then invested in a risk free collateral, whose interest is usually modified via a swap contract. The protection seller uses the interest generated by the default free collateral, together with the premium received by the protection buyer, both to respect the notes' holders rights and to ensure to the originator the protection agreed.

Differently from a fully funded synthetic CDO, in a unfunded synthetic CDO the originator transfers the reference portfolio stipulating more CDSs with the SPV. The CDSs differ for the risk they reflect. Furthermore, in this deal the investors are not supposed to response to any initial investments, so no notes are issued and consequently no risk free collateral is present.

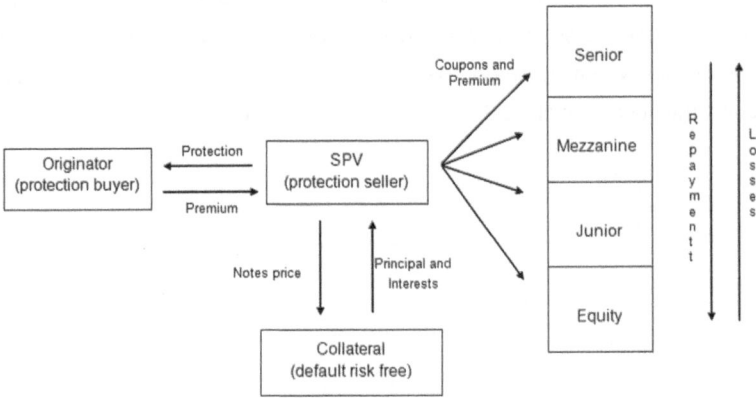

Figure 2: Fully funded synthetic CDO [3]

A *partially funded synthetic CDO* is the most common structure of a synthetic CDO, which mix characteristics of a totally funded CDO and unfunded CDO. In this contract, as represented in **Figure 3**, the credit risk is transferred using both CDS and Credit Linked Notes (CLN). Usually the originator transfers the credit to two counterparties.

A first counterparty (protection seller) sells protection to the originator for a super senior and a junior tranches of the reference portfolio, through two different CDS. These two pieces constitutes the unfunded part.

A second counterparty, usually a SPV, sells protection to the originator for the remaining volume of the reference portfolio. As in a fully funded CDO, the SPV has to invest in a risk free collateral, to guarantee the payments due to the originator. If a credit event in the reference portfolio occurs, the SPV can cover the losses selling a part of its collateral securities. The collateral securities are bought, with the money which SPV collected issuing CLN. The notes linked to the reference portfolio, are divided in tranches reflecting different classes of risk and their interest are paid back with the spreads that the originator, as protection buyer, owes to the SPV.

In the occurrence of a credit event, the junior CDS counterparty is the first to cover the losses.

---

[3] Bluhm, Overbeck and Wagner (2003).

When the cumulated losses of the reference portfolio exceed the upper limit of the junior piece, the notes' investors have to cover the losses according to the tranches seniority. Finally, when losses exceed the upper limit of the super senior tranches, the super senior CDS counterparty has to pay too.

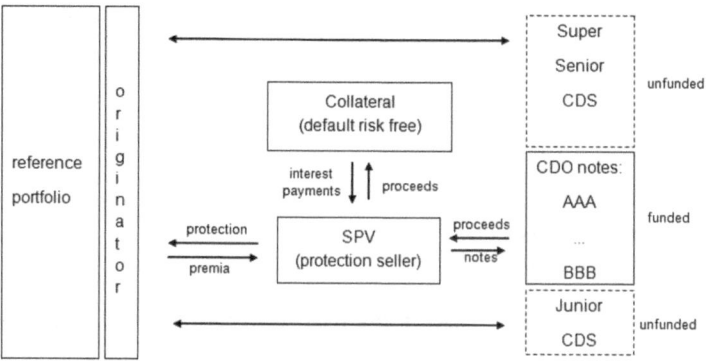

Figure 3: Partially funded synthetic CDO [4]

2.7 Credit default swap index

2.7.1 Definition

Credit default swap index are tradable products that allow investors to sell or buy protection on specific credit markets, through establishment of long or short position on the index.

These indices are standardized and global products. The most important are: CDX indices in North America and Emerging Markets, and the iTraxx indices for Europe and Asia. A more clear overview of the global indices follows in **Figure 4**.

As well as an index, CDS indices reflect the performance of a basket of assets, which are in this case single CDSs. CDX and iTraxx are both characterized by a basket of 125 individual CDS with equal weights inside the portfolio. When a credit event occurs, the name is immediately removed from the reference portfolio.

---

[4] Bluhm, Overbeck and Wagner (2003) p. 299.

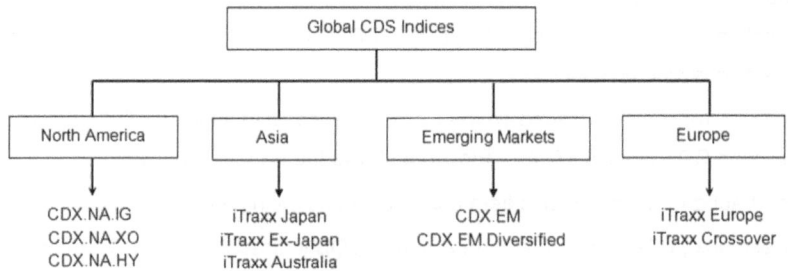

Figure 4: Overview of global CDS indices [5]

Each CDX index acts as standard CDS with a fixed portfolio of credits and a fixed annual coupon, divided into quarterly payments.

These indices, such as each instruments traded in the market, have a price called the market spread, which is determined by supply and demand. For this reason, the difference between the fixed coupon and the current price, has to be offset via an upfront.

In particular, if the fixed coupon is greater than the market spread, then the protection seller (long position) has to compensate an upfront to the protection buyer (short position).

2.7.2 Synthetic CDS index tranches

A tranche allows division of the total risk of a reference portfolio into several classes, characterized by different risk-return profiles. The greater the risk an investor decides to bear, the greater its return compensation.

As a CDS provides credit risk protection on an individual name and as CDS index, provides credit risk protection on a reference portfolio of several individual CDSs, a *tranche* CDS index provides credit risk protection on a particular amount of loss, of a reference portfolio constituted by several individual CDSs.

---

[5] Merrill Lynch (2006) p. 73.

A tranches is used to agree the specific pieces of the total losses, on which the protection is bought or sold. As in a CDS contract, the cost of the tranche protection is paid as coupon.

Standard tranches are traded on the North America CDS indices, CDX, and on the European CDS indices, iTraxx. **Figure 5** shows the tranches available.

Each traded tranched index differs according to its main characteristics.

For example, referring to the North America CDS indices market, CDX Investment Grade (CDX.IG) are divided into 0-3%, 3-7%, 7-10%, 10-15%, 15-30% and 30-100% tranches where the 0-3% class is called the equity tranches. Instead in the CDX High Yeld (CDX.HY) the tranches are the 0-10%, 10-15%, 15-25%, 25-35%, 35-100% , where the 0-10% and 10-15% represent the equity. Another difference between these two index lies in the way equity is traded. In CDX.IG the equity premium is the sum of an upfront and a 500 basis point (bp) spread, while in CDX.HY equity premium is only the upfront. The equity tranche trades at the 500 bp in both indices.

In the European landscape the iTraxx, except for the tranches width, is similar to the CDX. The tranches are into 0-3%, 3-6%, 6-9%, 9-12%, 12-22% and 22-100% where the 0-3% is the equity tranche.

	USA		Europe
	CDX.IG	CDX.HY	iTraxx
Maturity	3y,5y,7y,10y	3y,5y,7y,10y	3y,5y,7y,10y
Tranches	0-3%	0-10%	0-3%
	3-7%	10-15%	3-6%
	7-10%	15-25%	6-9%
	10-15%	25-35%	9-12%
	15-30%	35-100%	12-22%
	30-100%		22-100%

Source: JPMorgan

Figure 5: Summary of the available tranches index [6]

2.7.3 Synthetic tranches target

The synthetics tranches are instruments which allow investors to receive default protection, leverage exposure, hedging and trading opportunities.

---

[6] Source: JPMorgan (2006) p. 138.

16

When an investor decides to buy protection on a specific tranche, the protection is ensured only for the amount of losses which fall into that specific tranches.

That is, buying protection on an equity tranche ensures receiving money up to the attachment point. For this reason, buying protection through an equity tranche is cheaper than being long on CDS indices.

Tranches provide two kind of leverage exposure: one referred to the risk of portfolio losses and one referred to moves in the spread of the underlying portfolio.

The leverage exposure to the risk of portfolio losses can be explained as follows. The protection seller of the CDX.IG equity tranche and the protection seller of the CDX.IG index will receive the same annual amounts, equal to 500 bp, in case of no default. In case of one credit event the protection seller of the equity tranche will lose 16% of the notional, while the protection seller of the CDX.IG index will lose only 0.48%.[7]

The leverage exposure to spread moves, refers to the fact that synthetic CDO, being traded quickly and in relation of huge amounts, influence CDS spreads. The sensitivity of the tranche to the spreads is called "Delta" and is quoted too.

The equity tranches is the one with the highest sensitivity, i.e. the highest Delta. Being the first losses covering to the equity tranches, the spread paid/received by the protection buyer/seller will be the highest. The higher the seniority of the tranche, the lower the referred spread. This is why the delta decrease as the seniority increases.

The synthetic tranches became popular also for their use as hedge tool, against portfolio losses and spread moves in the underlying portfolio.

Investor wanting to hedge the credit risk of a portfolio of CDS can use synthetic tranches, which represent an alternative cheaper way to the CDS indices.

2.8   CDO Pricing: a general approach

2.8.1 Loss Distribution

Consider a static CDO whose reference portfolio is constituted by credit default swaps. The CDO investors are the protection sellers, which offer protection when

---

[7] Assuming a 60% loss on a credit , equal to assume a loss of 0.48% on 125 names portfolio (1/125x60%=0.48%). The protection seller of equity will pay 16% (0.48%/3%=16%) of the notional.

a credit event occurs, in return of a premium. A tranche is characterized by lower and upper bounds, called respectively L and U. Each tranche absorbs exclusively part of the cumulative loss percentage which falls between the attachment and the detachment point.

Let $L(t)$ be the cumulative loss on the reference portfolio and $M(t)$ the cumulative loss on a given tranche, then:

$$M(t) = \begin{cases} 0 & if \ L(t) \leq L \\ L(t) - L & if \ L \leq L(t) \leq U \\ U - L & if \ L(t) \geq U \end{cases}$$

Determining the value of the cumulative loss is fundamental to determine the payment that is due to the protection buyer by the protection seller. That is fundamental to determine the cash flow between these two counterparties and finally the CDO tranche price.

Consider a reference portfolio with n obligors having a notional amount A_i and a recovery rate R_n, with $n = 1,2,...,N$. Define the loss given default of the nth-obligor as $L_n = (1 - R_n)A_n$. Let τ_i denotes the time default of the nth-obligor and $N_n(t) = 1_{(\tau_n < t)}$. Now we are able to express the cumulative loss of the reference portfolio at time t as:

$$L(t) = \sum_{n=1}^{N} L_n N_n(t) \tag{2.1}$$

which is a jump process, where to each jump of $M(t)$ corresponds a cash flow transfer from the investor, that is the protection seller, to the originator, that is the protection buyer.

Assuming each obligor has the same notional amount and the same recovery rate, we can write the expected percentage loss of a given tranche.

In a discrete time setting we have:

$$EL_{[L,U]}(t_i) = \frac{\mathbf{E}[M(t_i)]}{U - L}$$
$$= \frac{1}{U - L} \sum_{n=1}^{N} \min[\max[L_n(t_i) - L], U - L] \tag{2.2}$$

In a continuous setting, assuming to know, P_L, the probability density of the distribution L, we can write the expected loss of a CDO tranche $[L, U]$ as:

$$EL_{[L,U]}(t_i) = \frac{1}{U-L} \int \min[\max[L_n(t_i) - L], U - L] dP_L(x)$$ (2.3)

The expected tranche loss can be written as[8]:

$$EL_{[L,U]}(t_i) = \frac{1}{U-L} \left(\int_L^1 (x - L)\, dP_L(x) - \int_U^1 (x - U)\, dP_L(x) \right).$$ (2.4)

In order to determine the tranches' premiums, which should reflect the expected loss of each class, the distribution losses of the portfolio $P_L(x)$ is required. However, this is where the most of the difficulties lie in.

The loss distribution of a portfolio constituted by a set of defaultable instruments depends on the probability of default of each asset, the effective loss in case of default, i.e. loss given default (LGD) and the effective impact a default of a firm causes on another firm. The latter issue, named default correlation, plays an important role on the determination of the loss distribution.

The literature about the CDO pricing model is based on two main approaches called respectively structural models and reduced form models.

Structural models use the firm's fundamental financial variables, such as asset and liabilities to determine the default time. In particular, following the first structural model by Merton (1974), the default happens when equity is below the debt. Conversely, reduced form models do not consider the link between the default and the economic situation described by the balance sheet. These models use market data as the only source of information, from which easily both default probability and credit risk dependencies are bootstrapped. Obviously these are easier to calibrate, but they suffer the lack of a real linking between the default probability and the fundamental variables.

In the next chapter several models aiming to obtain the loss distribution are reported.

---

[8] For a proof see Anna Schlosser (2010) p. 98.

2.8.2 Spread

The fair premium due to the protection seller by the protection buyer, can be determined using the same idea at the basis of a CDS contract, i.e. setting the fair premium W such that the present value of the default leg is equal to the present value of the premium leg. The default leg (DL) is the sum of all the expected losses of a tranche and the premium leg (PL) is the sum of the expected premia the protection seller receives. Assuming a constant annual spread for a given tranche we can write the two present value legs as follow:

$$DL = \sum_{i=1}^{k} \left(EL_{[L,U]}(t_i) - EL_{[L,U]}(t_{i-1}) \right) \cdot B(0,t_i) \tag{2.5}$$

$$PL = \sum_{i=1}^{k} \Delta_{i-1,i} \cdot W \cdot \left(1 - EL_{[L,U]}(t_i) \right) \cdot B(0,t_i) \tag{2.6}$$

where $B(0,t_i)$ is the discounting factor for each maturity t_i and k is the number of maturities.

The fair spread is obtained setting a premium W by equaling the present values of the two legs. That is:

$$W = \frac{\sum_{i=1}^{k} \left(EL_{[L,U]}(t_i) - EL_{[L,U]}(t_{i-1}) \right) \cdot B(0,t_i)}{\sum_{i=1}^{k} \Delta_{i-1,i} \cdot \left(1 - EL_{[L,U]}(t_i) \right) \cdot B(0,t_i)} \tag{2.7}$$

Being W constant by the previous assumption and t_i fixed in advance, the only unknown variable is the expected loss of the tranches. In order to compute it we have to model a default process for each obligor, taking into account the default correlation.

3   Credit Risk Modeling

This chapter will present an overview of the credit risk modeling literature. It will present the general setup which the most common models share and the derivation of the models themselves. This overview contains the models, which will be in the centre of the empirical part of the next chapters.

3.1   The Bernoulli Model

Assume the existence of a portfolio constituted by m counterparty and denote L_i the loss of the ith- obligor, with $i = 1, ..., m$.

Let $L = (L_1, ... , L_m)$ be a vector of random variables, whose marginal distributions $L_i \sim B(1; p_i)$ are Bernoulli. A two-state of world is assumed, where the obligors default with probability p_i, when $L_i = 1$, and survive with probability $1 - p_i$, when $L_i = 0$. That is:

$$L_i \sim B(1; p_i), \qquad L_i = \begin{cases} 1 & with\ p_i \\ 0 & with\ 1 - p_i \end{cases}$$

where $p_i = P[L_i = 1]$ is the probability of default.

Being the portfolio a set of m credit-risky assets, the portfolio loss is then a random variable L, defined as

$$L = \sum_{i=1}^{m} L_i.$$

The simplest situation corresponds to assume uniform default probability and lack of dependence between obligors. According to these assumptions, it is possible to write:

$$L_i \sim B(1; p) \quad and \quad L_i\ independent\ \forall\ i = 0, ... m.$$

Assuming a uniform default probability and the lack of default correlation, the portfolio loss $L = \sum_{i=1}^{m} L_i$ corresponds to a convolution of $i.i.d.$ Bernoulli variables that follow a binomial distribution $L \sim B(m; p)$.

According to the Binomial distribution properties the first and second moments are

$$E[L] = mp \quad and \quad Var[L] = mp(1-p).$$

Assuming obligors be still independent, but with different default probabilities lead to:

$$L_i \sim B(1; p_i) \quad and \quad L_i \ independent \ \forall \ i = 0, \dots m.$$

As before the portfolio loss is again a convolution but the first and second moments are now:

$$E[L] = \sum_{i=1}^{m} p_i \quad and \quad Var[L] = \sum_{i=1}^{m} p_i \, (1-p_i).$$

The assumption of independence simplifies the situation because it allows the use of the central limit theorem, thanks to which we could approximate the loss distribution as a Gaussian variable, at least for large portfolio.

The assumption of independence obviously simplifies the model, given the opportunity to recur to a close formula, due to use a Gaussian variable, instead of recur to a simulation via Monte Carlo.

The independence assumption in credit risk are useful as well they are so far away from the reality. For this reason modeling credit risk has to take in consideration the modeling of the *correlation* issue. This challenge is presented in the following models.

3.2 A Bernoulli mixture Model

3.2.1 The general case

Assuming independency in credit risk model and leaving out correlation issue is not realistic.

In the Bernoulli Mixture Model the loss of a portfolio has marginal losses again distributed with a Bernoulli distribution, that is $L_i \sim B(1; P_i)$.

The difference is that now, the default probability are randomized in a correlated way. $P = (P_1, \dots, P_m)$ is the random vector distributed following a distribution F supported in $[0,1]^m$, with realization vector $p = (p_1, \dots, p_m)$.

Furthermore, it is assumed that conditionally to a realization of the default probability the Bernoulli variables L_i are independent. Mathematically speaking:

$$L_i|P_i = p_i \sim B(1; p_i) \quad with \ L_i|P_i = p_i \ independent$$

The unconditional joint distribution of the L_i is:

$$P[L_1 = l_1, \dots, L_m = l_m] = \int_{[0,1]^m} \prod_{i=1}^m p_i^{l_i} (1 - p_i)^{1-l_i} \, dF(p_1, \dots, p_m) \qquad (3.1)$$

Each single loss is characterized by the following first and second moments:

$$E[L_i] = E[p_i] \quad and \quad Var[L_i] = E[p_i](1 - E[p_i]). \qquad (3.2)$$

The covariance between two losses L_i, L_j is:

$$Cov[L_i, L_j] = E[L_iL_j] - E[L_i]E[L_j] = Cov[P_i, P_j] \qquad (3.3)$$

and the correlation between default is:

$$Corr[L_i, L_j] = \frac{Cov[P_i, P_j]}{\sqrt{E[p_i][1 - E[p_i]]}\sqrt{E[p_j]\left[1 - E[p_j]\right]}}. \qquad (3.4)$$

3.2.2 Uniform default probability and uniform correlation case

A common simplification of the Mixture Model is to consider uniform default probability and uniform correlation. This is usually assumed for portfolios characterized by the same exposure, both in risk and size term. Assuming uniformity imply exchangeability[9] of the Bernoulli variables $L_i \sim B(1; P)$, where the random default probability is again distributed following the distribution F supported in $[0,1]^m$. Once assuming, as in the general case, conditional independence of the L_i, it is possible determine the joint distribution as:

---

[9] A random vector is said to be exchangeable if its joint distribution is symmetric under any permutation of the random vector; i.e. $(L_1, \dots, L_m) \sim (M_1, \dots, M_m)$ where $(M_1, \dots, M_m)$ is one of its possible permutation.

$$P[L_1 = l_1, \dots, L_m = l_m] = \int_0^1 p^k (1-p)^{m-k} \, dF(p) \tag{3.5}$$

$$\text{where} \quad k = \sum_{i=1}^{m} l_i \quad \text{and} \quad l_i \in \{0,1\}$$

The probability of having exactly k default is:

$$P[L = k] = \binom{m}{k} \int_0^1 p^k (1-p)^{m-k} \, dF(p) \tag{3.6}$$

3.3 Moody's KMV's and RiskMetrics' Model Approach

The model presented in this section belong to the family of threshold models, based on the idea that the default occurs when a relevant random variable lies below some critical barrier. According to the model the relevant random variable varies. The main idea is that in a given point of time, a firm defaults if its asset value is smaller than a certain threshold. Consider, as usual the existence of m counterparties, each of them characterized, at time T, by an asset value $A^{(i)}{}_T$. Each company is supposed have a critical threshold C_i, such that the firms default in the period $[0, T]$ if and only if at the end of this period, so when $t = T$, $A^{(i)}{}_T < C_i$.

Applying to this specific environment the previous Bernoulli Mixture Model, we can define each loss statistic as a Bernoulli distribution with parameter 1 and $P[A^{(i)}{}_T < C_i]$, that is:

$$L_i = \mathbf{1}_{(A^{(i)}{}_T < C_i)} \sim B\big(1; P[A^{(i)}{}_T < C_i]\big) \qquad (i = 1, \dots, m) \tag{3.7}$$

$A^{(i)}{}_T$ is the only random variable which determines the default event and in both models, it is assumed to be a process driven by underlying factors reflecting country and sector events. The main idea is modeling the default dependence between obligors via correlation between returns. One way is represented by building a correlation matrix using historical returns. However it is not the most practical way, especially in large portfolio case. Another one is to write the returns

24

as a sum of common factors and a firm specific factor, and to express the dependence according to different exposures to same common factor.

According to the Merton's model, the *Global Correlation Model*™ defines the asset log-return $r_i = \frac{A^{(i)}{}_T}{A^{(i)}{}_0}$ represented as follows:

$$r_i = R_i Y_i + \varepsilon_i \qquad\qquad where\ r_i \sim N(0,1). \qquad (3.8)$$

This model gives the possibility to model default correlation in a more direct way. In fact here it is assumed that the asset log return is a sum of two parts. The Y_i common to all the asset return, called composite factor, and the other one ε_i, that is the specific part, called idiosyncratic effects. The composite factor is the sum of country and factor indices.

The coefficient R_i determines how the ith- firm is influenced by the common factor and comparing these coefficient it is possible to obtain an idea of how different asset are correlated.

The idiosyncratic effects of each firm are assumed be independent between them and independent of the composite factors.

Both model assume that the asset value log-returns are normally distributed, that is:

$$r_i \sim N(0,1), \qquad\qquad Y_i \sim N(0,1), \qquad\qquad and\ \varepsilon_i \sim N(0,1 - R_i{}^2)$$

Given that the driving variable $A^{(i)}{}_T$ is replaced by the asset log-return r_i, the threshold C_i is replaced by the corresponding c_i. Therefore the loss statistic can rewritten as:

$$L_i = \mathbf{1}_{(r_i < c_i)} \sim B(1; P[(r_i < c_i)]) \qquad\qquad (i = 1, ..., m) \qquad\qquad (3.9)$$

By (3.5) it is possible rewrite the default condition as:

$$\varepsilon_i < c_i - R_i Y_i \qquad\qquad (i = 1, ..., m) \qquad\qquad (3.10)$$

Let $p_i = P[(r_i < c_i)]$ be the default probability of the ith- obligor, being $r_i \sim N(0,1)$, it is immediately possible to obtain:

$$c_i = N^{-1}[p_i] \qquad (i = 1, ..., m) \qquad (3.11)$$

Once divided the idiosyncratic term by its variance, in order to scale it, and substituting the threshold with the (3.11), it is possible to obtain the normal random variable:

$$\tilde{\varepsilon}_i = \frac{N^{-1}[p_i] - R_i Y_i}{\sqrt{1 - R_i^2}} \qquad (3.12)$$

Given that in both models the horizon is $T = 1$ year, the one year probability of default for the ith-obligor, conditional on the factor Y_i, is given by:

$$p_i(Y_i) = N\left[\frac{N^{-1}[p_i] - R_i Y_i}{\sqrt{1 - R_i^2}}\right] \qquad (i = 1, ..., m) \qquad (3.13)$$

The only stochastic term is the composite factor Y_i and conditional on $Y_i = z$ the condional one-year probability of default is given therefore by:

$$p_i(z) = N\left[\frac{N^{-1}[p_i] - R_i z}{\sqrt{1 - R_i^2}}\right] \qquad (i = 1, ..., m) \qquad (3.14)$$

Given the (3.9) we follow the same elaboration of the Bernoulli mixture model framework. Assuming again that $l_i \in \{0,1\}$, as in (3.1) we specify the joint probability:

$$P[L_1 = l_1, ..., L_m = l_m] = \int_{[0,1]^m} \prod_{i=1}^{m} q_i^{l_i} (1 - q_i)^{1-l_i} \, dF(q_1, ..., q_m) \qquad (3.15)$$

where the distribution F is now given by:

$$F(q_1, \dots, q_m) = N_m[p_1^{-1}(q_1), \dots p_m^{-1}(q_m); \Gamma] \qquad (3.16)$$

where, $N_m[\dots; \Gamma]$ is a cumulative multivariate normal distribution with correlation matrix Γ. $\Gamma = (\varrho_{ij})_{1 \le i,j \le m}$ is the asset correlation of the asset log-return.

Notice that when the composite factors are expressed through a weighted sum of country and sector indices $(\alpha_j)_{j=1,\dots,J}$, such that:

$$Y_i = \sum_{j=1}^{J} w_{ij}\alpha_j \qquad (3.17)$$

then the conditional probability of default is given by:

$$p_i(Y_i) = N\left[\frac{N^{-1}[p_i] - R_i(w_{i1}\alpha_1 + \dots + w_{iJ}\alpha_J)}{\sqrt{1 - R_i^2}} \right] \qquad (3.18)$$

3.4 One-Factor Model

The one-factor model relies on both Moody's KMV and RiskMetrics context. However, it differs from them because assume the existence of a one-factor common to all obligors, therefore assuming uniform asset correlation. The (3.3) in the light of the model specification, is hereby rewritten as:

$$r_i = \sqrt{\varrho}Y + \sqrt{1 - \varrho}Z_i \qquad (i = 1, \dots, m) \qquad (3.19)$$

where the previous composite factor Y_i is now replaced by a one-factor common to all obligors, such that $Y \sim N(0,1)$. Moreover, ϱ is the uniform log-return correlation, unique for all obligors, and $Z_i \sim N(0,1)$ is the idiosyncratic term. Again, Z_i is assumed be independent of the one-factor Y.

Under new model specification the default probability can be rewritten as:

$$p_i(Y) = N\left[\frac{N^{-1}[p_i] - \sqrt{\varrho}Y}{\sqrt{1-\varrho}}\right] \qquad (i = 1, ..., m) \qquad (3.20)$$

and the default probability conditional to $Y = z$ is therefore:

$$p_i(z) = N\left[\frac{N^{-1}[p_i] - \sqrt{\varrho}z}{\sqrt{1-\varrho}}\right] \qquad (i = 1, ..., m) \qquad (3.21)$$

In **Figure 6** is possible to observe how the conditional default probability change at changing in the realizations of the factor. The default probability of default of the ith- obligors is assumed to be 30 bp and the uniform correlation $\varrho = 20\%$.

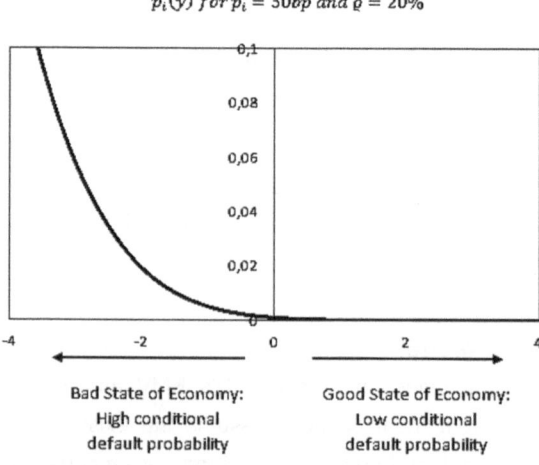

Figure 6: Asset Value One-Factor Model: Conditional default probability as a function of the factor realizations Y=y [10]

In **Figure 7** it is possible to observe how the conditional default probability as function of the default probability. Assuming the uniform correlation $\varrho = 20\%$, the figure shows three states of the economy. In particular are considered the specific case of the factor realizations Y=-3, 0, 3.

---

[10] Bluhm, Overbeck and Wagner (2003) p. 81.

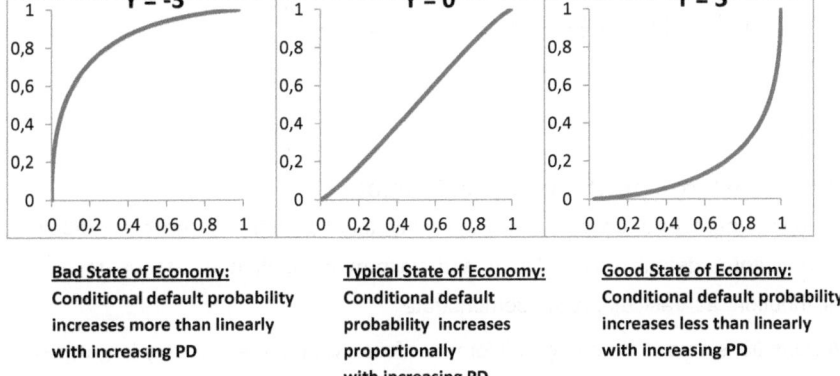

$p_i(y)$ *in dependence on* p_i, *for fixed* $\varrho = 20\%$ *and* $Y = -3, 0, 3$

Bad State of Economy:
Conditional default probability increases more than linearly with increasing PD

Typical State of Economy:
Conditional default probability increases proportionally with increasing PD

Good State of Economy:
Conditional default probability increases less than linearly with increasing PD

Figure 7: Asset Value One-Factor Model: Conditional default probability as a function of the average 1-year default probability p_i [11]

The probability of having exactly i default is, as before, an average of the i default probabilities, conditional to $Y = z$, averaged over all the possible Y realizations and weighted with the probability density function g(z):

$$P[L = i] = \int_{-\infty}^{\infty} P[L = i | Y = z]\, g(z) dz \qquad (3.22)$$

where the probability of having exactly i default, conditional to $Y = z$ is given by the binomial distribution:

$$P[L = i | Y = z] = \binom{m}{i} p(z)^i (1 - p(z))^{m-i}. \qquad (3.23)$$

Substituting (3.23) and the (3.21) in (3. 22) leads to:

$$P[L = i] = \int_{-\infty}^{\infty} \binom{m}{i} \left[N \left(\frac{N^{-1}[p_i] - \sqrt{\varrho}z}{\sqrt{1-\varrho}} \right) \right]^i \left[-N \left(\frac{N^{-1}[p_i] - \sqrt{\varrho}z}{\sqrt{1-\varrho}} \right) \right]^{m-i} g(z) dz \qquad (3.24)$$

---

[11] Bluhm, Overbeck and Wagner (2003) p. 82.

Finally, the distribution function of the default is:

$$P[L \leq k] =$$

$$\sum_{i=0}^{k} \binom{m}{i} \int_{-\infty}^{\infty} \left[N \left(\frac{N^{-1}[p_i] - \sqrt{\varrho}z}{\sqrt{1-\varrho}} \right) \right]^i \left[1 - N \left(\frac{N^{-1}[p_i] - \sqrt{\varrho}z}{\sqrt{1-\varrho}} \right) \right]^{m-i} g(z) dz \tag{3.25}$$

3.5   The large portfolio approximation

When the number of obligors tends to infinity, the distribution for uniform portfolios, with uniform default probability p and uniform correlation ϱ, tends to a limit distribution, as Vasicek (1987) demonstrate.

A portfolio is said to be a large uniform portfolio, when there is a very large $m \to \infty$ number of obligors having uniform exposure amount.

Once introduced as usual a Bernoulli mixture model, each obligor is described by a loss statistic:

$$L_i = \mathbf{1}_{(r_i < c_i)} \sim B(1; P[(r_i < c_i)]) \qquad (i = 1, ..., m). \tag{3.26}$$

The term $r_i \sim N(0,1)$ follows a one-factor process of the kind:

$$r_i = \sqrt{\varrho} Y + \sqrt{1 - \varrho} Z_i \qquad (i = 1, ..., m) \tag{3.27}$$

and, being default probability assumed uniform, the threshold is given by

$$c_i = N^{-1}[p] \qquad (i = 1, ..., m). \tag{3.28}$$

Let be $g_{p,\varrho}(y)$ the default probability conditional on the realizations of the common factor, as usual, it is given by:

$$g_{p,\varrho}(y) = N \left[\frac{N^{-1}[p] - \sqrt{\varrho}y}{\sqrt{1 - \varrho}} \right] \qquad (i = 1, ..., m) \tag{3.29}$$

Vasicek proved that in such context the percentage portfolio loss distribution

$$L_m^\% = \frac{1}{m} \sum_{i=1}^{m} \mathbf{1}_{\{r_i < N^{-1}[p]\}}$$

(3.30)

admits almost surely a limit for $m \to \infty$ as follows

$$P\left[\lim_{m \to \infty} \left(L_m^\% - g_{p,\varrho}(y)\right) = 0\right] = 1$$

(3.31)

So almost surely, under the previous assumption the percentage portfolio loss distribution can be replaced by the default conditional on the realization of the common factor. That is, under the Vasicek's limit theorem:

$$F_{p,\varrho}(x) := P[L_m^\% < x] = P\left[g_{p,\varrho}(y) < x\right]$$

$$= N\left[\frac{(N^{-1}[x]\sqrt{1-\varrho}) - N^{-1}[p]}{\sqrt{\varrho}}\right]$$

(3.32)

Computing the derivative of the distribution function with respect to x it is possible to recover the corresponding density function $f(x)$:

$$f_{p,\varrho}(x) = \frac{\partial F_{p,\varrho}(x)}{\partial x}$$

$$= \sqrt{\frac{1-\varrho}{\varrho}} \exp\left\{\frac{1}{2}(N^{-1}(x))^2 - \frac{1}{2\varrho}\left(N^{-1}(p) - N^{-1}[x]\sqrt{1-\varrho}\right)^2\right\}$$

(3.33)

The **Figure 8** shows how the density $f_{p,\varrho}(x)$ changes, varying the parameters p and ϱ.

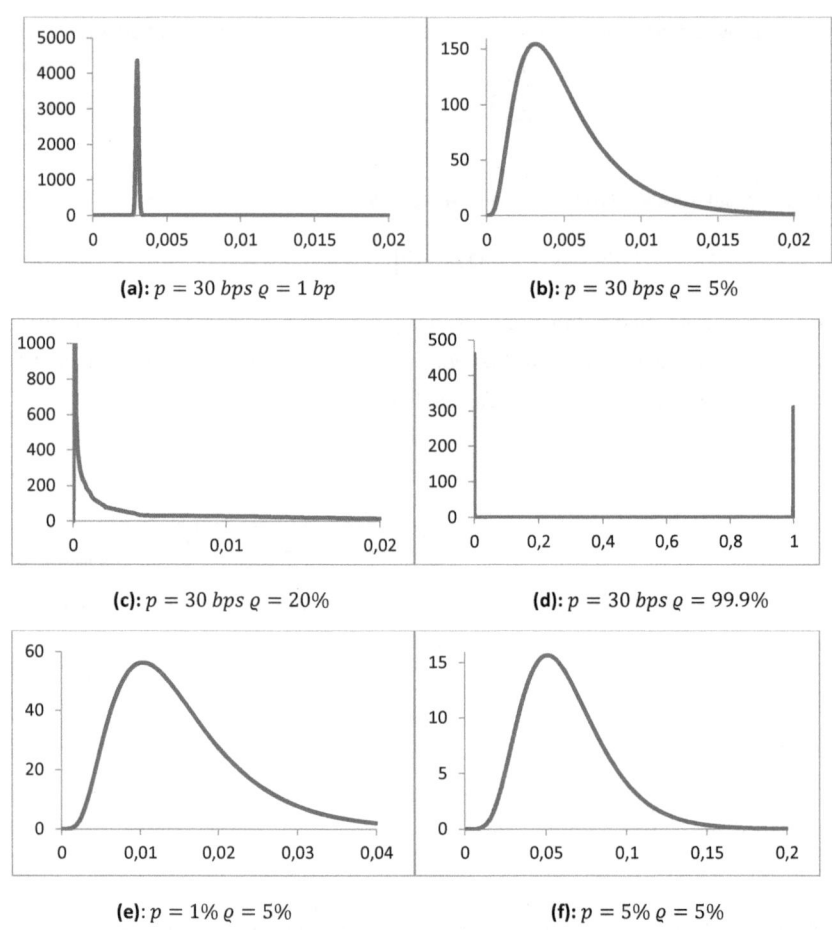

(a): $p = 30 \; bps \; \varrho = 1 \; bp$

(b): $p = 30 \; bps \; \varrho = 5\%$

(c): $p = 30 \; bps \; \varrho = 20\%$

(d): $p = 30 \; bps \; \varrho = 99.9\%$

(e): $p = 1\% \; \varrho = 5\%$

(f): $p = 5\% \; \varrho = 5\%$

Figure 8: The probability density $f_{p,\varrho}(x)$ for different parameters scenarios p and ϱ [12]

In particular it is possible to notice the four extreme cases admitted by the loss density according to the four extreme value taken by the parameters p and ϱ.

- Absence of correlation case, $\varrho = 0$:

   In this situation, the (3.27) is simply $r_i = Z_i$, and the loss variable becomes

$$L_i = \mathbf{1}_{\left(Z_i < N^{-1}(p)\right)} \sim B(1; p).$$

---

[12] Bluhm, Overbeck and Wagner (2003) p. 89.

In a correlation-free situation the portfolio loss $L_m = \sum_{i=1}^{m} L_i$ tends to a binomial distribution, $L_m \sim B(m; mp)$. In addition the percentage loss defined as $L^{(m)} = \sum_{i=1}^{m} w_i LGD_i L_i$, converges to p by the Law of Large Numbers.

So the density function $f_{p,0}(x)$ is the density of a degenerate function concentrate in p. Panel (a) of **Figure 8** shows the loss density in a situation of almost null correlation $\varrho = 1bp$ with $p = 30\ bps$.

- Perfect correlation case, $\varrho = 1$:

   Perfect correlation means that default of one obligor imply almost surely default of all the other obligors of the portfolio. For this reason the percentage portfolio loss is independent on the number of obligors m and follows a binomial distribution $B(1; p)$, where $P[L = 1] = p$ and $P[L = 0] = 1 - p$. As illustrated in panel (d) of the **Figure 8** it is possible to notice that the loss density is concentrated in only two points, meaning that the loss portfolio could be total or null. There are no intermediary situation and this case is called the "all or nothing" loss case.

- $p = 0$:

   In this situation where the uniform default probability is assumed to be 0, all the obligors survive almost surely. That is $P[L = 0] = 1 - p = 1$.

- $p = 1$:

   In this situation where the uniform default probability is assumed to be 1, all the obligors default almost surely. That is $P[L = 0] = 1 - p = 0$.

3.6   Multifactor models

The multi-factor model is an extension of the previous sections which consider more than a unique driving common factor. Mathematically, we can decompose the asset return of the ith- obligor as:

$$X_i = \beta_i Y_i + \hat{\varepsilon}_i \tag{3.34}$$

where Y_i is the composite factor of the ith- obligor and it is a weighted sum of indices:

$$Y_i = \sum_{n=1}^{N} \alpha_{i,n} Z_i \tag{3.35}$$

Following the Moody KMV separation approach, Z_i denoting the sector indices or the geographic indices are distributed as a standard Gaussian. Furthermore $\hat{\varepsilon}_i \sim N(0, 1 - R_i^2)$ where:

$$R_i^2 = \text{Var}\left[\beta_i \sum_{n=1}^{N} \alpha_{i,n} Z_i\right] = \beta_i^2 \sum_{k,l=1}^{N} \alpha_{k,n} \alpha_{l,n} \text{Cov}[Z_k Z_l], \tag{3.36}$$

and Y_i and $\hat{\varepsilon}_i$ are independent for any i.

Multiplying the idiosyncratic vector for its standard deviation and considering that $R_i^2 = \beta_i^2$ [13], we can rewrite the asset return of the ith- obligor as:

$$X_i = \beta_i Y_i + \sqrt{1 - \beta_i^2} \, \varepsilon_i$$

$$\tag{3.37}$$

where $\varepsilon_i \sim N(0,1)$ and again Y_i and ε_i are independent for any i.

---

[13] C.Bluhm and L.Overbeck (2007) p. 322.

4 Copula functions and dependency concepts

4.1 Copulas Basics

Modeling default of several obligors implies modeling the default probability of the single obligor as well as the dependence structure between obligors.

A general distribution function, in our example a distribution function of a portfolio of several obligors, contains information about both marginal obligor distribution and their correlation structure. However these two parts are implicit in it. A copula function is a tool, allowing a way of isolating the description of such dependence structure.

Further details on copula function and proofs of theorem can be found in Joe (1997), Nelsen (2006) and Cherubini *et al.* (2004).

Definition 4.1 (copula) *A I-dimension copula is a distribution function, $C:$ $[0,1]^I \rightarrow [0,1]$ with uniform marginal distribution satisfying the following 3 properties:*

(a) $C(u_1, \ldots, u_I)$ *is increasing in each component u_i.*

(b) $C(1, \ldots, 1, u_i, 1, \ldots 1) = u_i$ *for all $i \in \{0, \ldots, I\}$, $u_i \in [0,1]$;*

(c) *(rectangle inequality): for all $(a_1, \ldots, a_I), (b_1, \ldots, b_I) \in [0,1]^I$ with $a_i < b_i$ we have*

$$\sum_{i_1=1}^{2} \sum_{i_2=1}^{2} \cdots \sum_{i_I=1}^{2} (-1)^{i_1+i_2+\cdots+i_I} C(v_{i_1}, v_{i_2}, \ldots, v_{i_I}) \geq 0$$

where $v_{j_1} = a_j$ and $v_{j_2} = b_j$ for all $j = 1, \ldots, I$.

The first property is a necessary requisite for any multivariate distribution functions and the second is the equivalent mathematic way to require uniform marginal distribution. The third property ensures that for any random vector $(U_1, \ldots, U_I)'$ with distribution function C, $P[a_1 \leq U_1 \leq b_1, \ldots, a_I \leq U_I \leq b_I]$ is non-negative.

The following theorem shows the importance of copulas, stating that all multivariate distribution functions contain copulas, as well that copulas can be used to construct multivariate distribution functions, starting from the marginal distributions.

Theorem 4.2 (Sklar)   *Let $X_1, ..., X_I$ be random variables , $F_1, ..., F_I$ be their marginal distributions and F be the joint distribution. Then there exists a copula $C : [0,1]^I \rightarrow [0,1]$ such that for $\forall (x_1, ..., x_I) \in R^I$,*

$$F(x_1, ..., x_I) = C(F_1(x_1), ..., F_I(x_I) = C(F(x)),$$

if $F_1, ..., F_I$ are continuous then C is unique; conversely C is uniquely determined on $RanF_1 \cdot ... \cdot RanF_I$, where $RanF_i$ is the range of F_i. Moreover if C is a copula and $F_1, ..., F_I$ are univariate distribution functions then F is the joint distribution function with margins $F_1, ..., F_I$.

This is the fundamental theorem in copula framework, because ensuring that for any multivariate distribution, the univariate margins and the dependence structure can be separated, and the dependence structure is an implicit characteristic of the copula function.

Some more properties of copula functions follows.

Propositions 4.3

(*i*)   *(Invariance to increasing transformations.) Let $X = (X_1, ..., X_I)$ be a vector of random variables with copula $C(u)$. Then for any strictly increasing set of functions $f_i : R \rightarrow R$, $C(\cdot)$ is again a copula of the so built random vector, $X' = (f_1(X_1), ..., f_I(X_I))$.*

(*ii*)   *Let be $C : [0,1]^I \rightarrow [0,1]$ a copula. The copula C is non-decreasing in each argument. That is, if $v \in [0,1]^I$ then*

$$C(v) \leq C(v_{-j}, v'_j) \quad \forall 1 \geq v'_j > v_j, \ \forall j \leq I.$$

(*iii*)   *(Fréchet bounds.) For every $v \in [0,1]^I$ any I-dimensional copula C satisfies the following inequality:*

$$max \left\{ \sum_{i=1}^{I} v_i + 1 - I, 0 \right\} \leq C(v) \leq min(v_1, v_2, ..., v_I).$$

The Fréchet limits are important since they represent the upper and the lower bound to a copula. Moreover, the bounds represent the largest possible positive and negative dependence.

Remark:

The Fréchet bounds has been expressed in terms of copula but Fréchet bounds do exists for any multivariate distribution **F(x)** with margins $F_1, ..., F_I$:

$$max\left\{\sum_{i=1}^{I} F_i(x_i) + 1 - I, 0\right\} \leq \mathbf{F}(\mathbf{x}) \leq \min(F_1(x_1), \dots, F_i(x_i)).$$

4.2   Examples of copulas

Copulas can be divided into three main categories: *fundamental, implicit* and *explicit* copulas. *Fundamental* copulas are the ones expressing a particular dependence structure, as the product copula.

Implicit copulas are so called because they are an implicit consequence of the Sklar's theorem application to well known multivariate distribution, as the Gaussian and t-students multivariate distributions. Implicit copulas are Gaussian copula and t-copula. These do not have a simple closed expression.

Explicit copulas are constructed following mathematical steps and have simple closed expression; Archimedean copulas are examples of explicit copulas.

4.2.1 Product copula

This is the simplest example of a copula and corresponds to the uniform distribution on $[0,1]^I$.

The product copula, also called the independence copula, is given by:

$$\sqcap (v_1, v_2, \dots, v_I) = \prod_{i=!}^{I} v_i$$

As the name suggest it is the copula of independent random variables. From the Sklar theorem it is clear that random variables $X_1, X_2, \dots, X_I$ are independent if and only if the I-dimensional copula C these random variables is:

$$C\big(F_1(x_1), \dots, F_I(x_I)\big) = \sqcap \big(F_1(x_1), \dots, F_I(x_I)\big)$$

4.2.2 Gaussian Copula

Let $X = (X_1, X_2, \dots, X_I)$ be a vector of normally distributed random variables, such that $X \sim N_I(\boldsymbol{\mu}, \boldsymbol{\Sigma})$. Let $Z \sim N_I(\mathbf{0}, \boldsymbol{P})$ be the standard normal vector, such that $P = \wp(X)$ is the correlation matrix of X. Being standardization a transformation of strictly increasing function, the copula of X is the same of Z. This copula is given by:

$$C_P^{GA}(\boldsymbol{u}) = P(\Phi(Z_1) \leq u_1, ..., \Phi(Z_I) \leq u_I)$$

$$= \Phi_P(\Phi^{-1}(u_1), ..., \Phi^{-1}(u_I))$$

where $\Phi(Z_1) = P(Z_1 \leq z_1)$ and Φ_P refers to the joint distribution function of $\boldsymbol{Z}$.

The Gaussian copula does not have a closed form but can be expressed as an integral over the density of $\boldsymbol{Z}$. Consider the 2-dimenional copula $C_\rho^{GA}(u_1, u_2)$ where $\rho = \rho(X_1, X_2)$ is the correlation between the two random variable X_1 and X_2. Then for $|\rho| < 1$:

$$C_P^{GA}(u_1, u_2) = \int_{-\infty}^{\Phi^{-1}(u_1)} \int_{-\infty}^{\Phi^{-1}(u_I)} \frac{1}{2\pi(1-\rho^2)^{1/2}} exp\left\{ \frac{-(s_1^2 - 2\rho s_1 s_2 + s_2^2)}{2(1-\rho^2)} \right\} ds_1 ds_2.$$

Product copula can be seen as a particular case of the Gaussian copulas. A Gaussian copula with correlation matrix $P = I_I$, where I_I is a identity matrix of order I.

Gaussian copula make easy draw random sample from it. The following algorithm generates random sample from the Gaussian copula with correlation matrix P:

1. Generate the vector $\boldsymbol{Z} = (Z_1, Z_2, ..., Z_I)$ jointly distributed following $\boldsymbol{Z} \sim N_I(\boldsymbol{0}, \boldsymbol{P})$.

2. The vector $\boldsymbol{U} = (U_1, U_2, ..., U_I)$ with $U_i = \Phi(Z_i)$ is distributed according to $C_P^{GA}(\boldsymbol{u})$. A further characteristic of the Gaussian copulas is their flexibility, which allows to link random variable with a Gaussian copula, even if the marginal distribution are not normally distributed.

The **Figure 9** shows the density of a Gaussian Copula whit correlation $\rho = 60\%$.

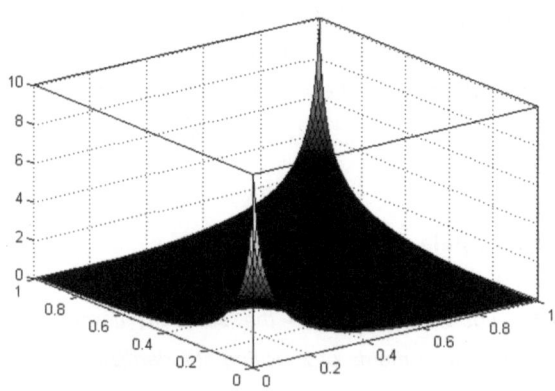

Figure 9: Density of a bivariate Gaussian copula whit correlation $\rho = 60\%$ [14]

4.2.3 t-Copula

Let $Z = (Z_1, Z_2, ..., Z_I)$ be a vector of normally distributed random variables, such that $Z \sim N_I(0, P)$. Let $Y = (Y_1, Y_2, ..., Y_I)$ be a vector of χ^2-variables, with v degrees of freedom, which are independent of Z.

Let $U = (U_1, U_2, ..., U_I)$ be a vector of random variables such that

$$U_i := t_v \left(\frac{\sqrt{v}}{\sqrt{Y}} Z_i \right),$$

then the distribution function $C_{v,P}^t(u)$ is a copula and it is called the t-copula with v degrees of freedom and correlation matrix P.

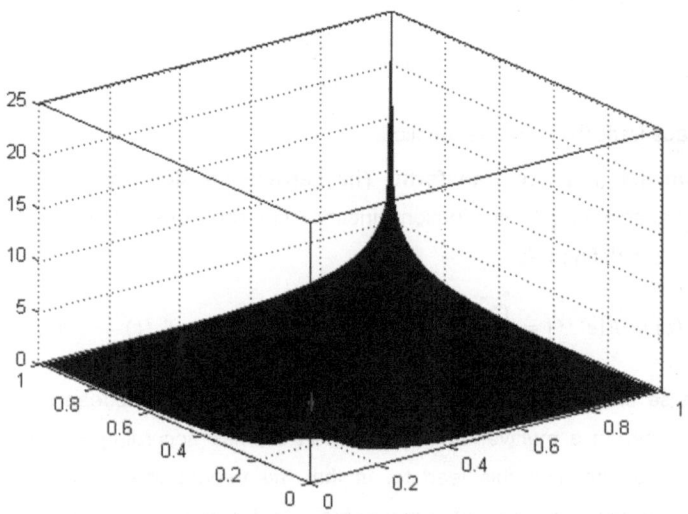

Figure 10: Density of a bivariate t-copula with correlation $\rho = 70\%$ and $v = 4$ [14]

4.2.4 Archimedean Copulas

Archimedean copulas are characterized by a very low number of free parameters, usually corresponding to one or two.

Archimedean copulas can be constructed using a function $\phi: [0,1] \rightarrow R_+$ called generator of the copula. The generator should be a continuous, decreasing and convex function such that $\phi(0) = +\infty$ and $\phi(1) = 0$.

---

[14] Own Source.

Pseudo-inverse of ϕ has to be defined too. In particular the *pseudo-inverse is the function:*

$$\phi^{[-1]}(v) = \begin{cases} \phi^{-1}(v), & 0 \leq v \leq \phi(0) \\ 0, & \phi(0) \leq v \leq +\infty \end{cases}$$

By definition the pseudo-inverse satisfies:

$$\phi^{[-1]}(\phi(v)) = v \qquad \forall v \in [0,1]$$

Once defined the generator and its pseudo-inverse, it is possible give a definition of Archimedean copula function.

An Archimedean copula function $C^A(x) : [0,1]^I \to [0,1]$ is a copula function that can be represented as follows:

$$C^A(x) = \phi^{[-1]}\left(\sum_{i=1}^{I} \phi(x_i)\right).$$

Let Laplace transformation be recalled:

Definition 4.4 (Laplace transform) The Laplace transform of a non-negative random variable Y, with distribution function $G(y)$ and density function $g(y)$ (if it does exist), is defined as:

$$\mathcal{L}_Y(t) := E[e^{-tY}] = \int_0^\infty e^{-tY} \, dG(y) = \int_0^\infty e^{-tY} g(y) dy =: \mathcal{L}_g(t), \qquad \forall t \geq 0.$$

The inverse of Laplace transforms is a generator of a Archimedean copula. It is easy to generate a multivariate Archimedean copulas according to the previous definition, but the limit this lead to, is that the dependence structure is then captured by only one or two parameters. So these are not sufficient if the dependency structure has to be modeled in more details.

The most important Archimedean copulas are following reported. For more completed list see Nelsen(1999).

Gumbel Copula

The Generator is given by $\phi(t) = (-\ln t)^\theta$, hence its inverse is $\phi^{[-1]}(t) = \exp(-t^{\frac{1}{\theta}})$. The Gumbel n-copula is therefore:

$$C(u_1, u_2, \dots, u_n) = \exp\left\{-\left[\sum_{i=1}^{n} (-\ln u_i)^\theta\right]^{\frac{1}{\theta}}\right\} \qquad \text{with } \theta \geq 1.$$

Clayton Copula

The Generator is given by $\phi(t) = (u^{-\theta} - 1)$, hence its inverse is $\phi^{[-1]}(t) = (1 + t)^{-\frac{1}{\theta}}$; The Clayton n-copula is therefore:

$$C(u_1, u_2, \ldots, u_n) = \left[\sum_{i=1}^{n} u_i^{-\theta} - n + 1\right]^{-\frac{1}{\theta}} \quad with\ \theta \geq 0\,.$$

Frank Copula

The Generator is given by $\phi(t) = ln\left(\frac{e^{-\theta t}-1}{e^{-\theta}-1}\right)$ hence its inverse is $\phi^{[-1]}(t) = -\frac{1}{\theta}\ln(1 - e^t(e^\theta - 1)$; the Frank n copula is given by:

$$C(u_1, u_2, \ldots, u_n) = -\frac{1}{\theta} ln\left\{1 + \frac{\prod_{i=1}^{n}(e^{-\theta u_i} - 1)}{(e^{-\theta} - 1)^{n-1}}\right\} with\ \theta \geq 1\,.$$

In **Table 1** the specifications of the generators of the most common Archimedean copulas are reported. In addition the inverse and the Laplace transforms, that is the distribution and the density function of the r.v. Y are reported as well.

Name	Gumbel	Clayton	Frank
$\phi(t)$	$\phi(t) = (-ln\ t)^\theta$	$\phi(t) = (u^{-\theta} - 1)$	$\phi(t) = ln\left(\dfrac{e^{-\theta t} - 1}{e^{-\theta} - 1}\right)$
$\phi^{[-1]}(t)$	$\phi^{[-1]}(t) = \exp(-t^{\frac{1}{\theta}})$	$\phi^{[-1]}(t) = (1 + t)^{-\frac{1}{\theta}}$	$\phi^{[-1]}(t) = -\dfrac{1}{\theta}\ln(1 - e^t(e^\theta - 1)$
θ	$\theta \geq 1$	$\theta \geq 0$	$\theta \geq 1$
$G(y)$	α-stable, $\alpha = (1/\theta)$	Gamma$(1/\theta)$	Logarithmic series on $\mathbb{N}_+$, with $\alpha = (1 - e^{-\theta})$
$g(y)$	(no closed form is known)	$\dfrac{1}{\Gamma(1/\theta)}e^{-y}y^{(1-\theta)/\theta}$	$P[Y = k] = \dfrac{-1}{\ln(1 - \alpha)}\dfrac{\alpha^k}{k}$

Table 1: Summary of some generators, their inverse and Laplace transforms for Gumbel, Clayton and Frank copulas [15]

---

[15] Marshall and Olkin (1988).

4.3 Tail dependence

Definition A bivariate copula $C(u, v)$ has a *upper tail dependence* with parameter λ_U if:

$$\lim_{u \to 1} \frac{1 + C(u, u) - 2u}{1 - u} = \lambda_U > 0.$$

C has a *lower tail dependence* with parameter λ_L if:

$$\lim_{u \to 0} \frac{C(u, u)}{u} = \lambda_L > 0.$$

The coefficients of upper and lower tail dependence are measures depending only on the copula of a pair of random variables X_1 and X_2 with continuous marginals F_1 and F_2. These coefficients express measures of *dependence* in the tails of a bivariate distribution.

In particular, *lower tail dependence* means that when $u \to 0$ the probability mass $C(u, u)$ tends to zero like $\lambda_L u$, and not like the area of the square u^2. That is, in the corner (0,0) of the square $[0, u]x[0, u]$ there must be a strong singularity of the copula's density.

Upper tail dependence means the same but in the corner (1,1) of the square $[0, u]x[0, u]$.

Roughly speaking, *upper tail dependence* means that there is a tendency of X_2 to assume extremes values when X_1 assume extremes values as well. Being copula functions used to model credit default dependence, tail dependence represent a huge problem when the random variable are financial losses.

By tail dependence definition[16] it is possible to obtain the upper and lower tail dependence coefficient λ_U and λ_L for the family of copulas described.

In particular, by applying the definition, it is possible demonstrating that the Gaussian copula has a tail dependence coefficient $\lambda = 0$ (provided the correlation be $\rho < 1$). That is, the Gaussian copula is asymptotically independent in both tails. Regardless of correlation we choose, if we go far enough into the tail, extreme events appear to occur independently in each margin.

Conversely, the t-copula presents tail dependence in both the tails. In panel (a) of **Figure 11** are represented 2000 samples from a Gaussian copula with correlation $\rho = 0.6$ and in panel (d) 2000 samples from a t-copula with correlation $\rho = 0.7$ and

---

[16] McNeil, Frey & Embrechts (2006) p. 209.

$v = 4$. It is possible to notice that in the t-copula graph samples cluster more in the lower and upper corner than the Gaussian graph.

The Archimedean copulas present an asymmetric dependence structure. In particular the Gumbel copula presents a upper tail dependence but not lower tail dependence, while conversely the Clayton copula has a lower dependence but not an upper dependence. Panel (b) and (c) of **Figure 11** shows respectively 2000 samples from a Gumbel with parameter $\theta = 2$ and a Clayton with parameter $\theta = 2.2$.

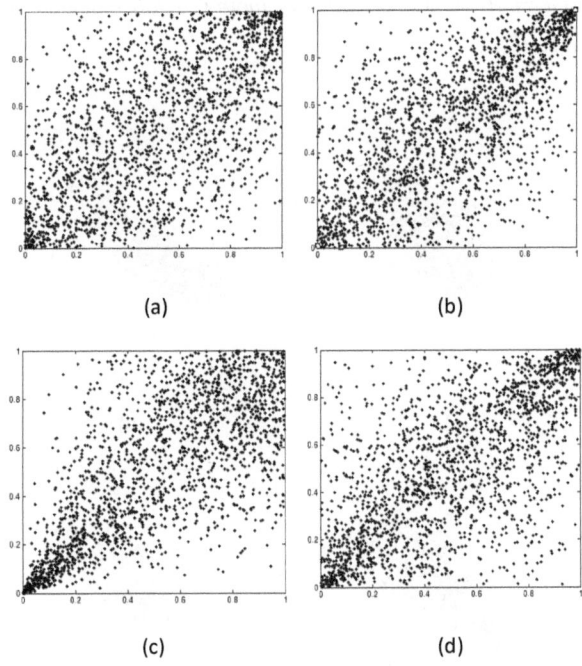

(a) (b)

(c) (d)

Figure 11: Two thousand simulated points from the (a) Gaussian, (b) Gumbel, (c) Clayton and (d) t copulas [17]

The Sklar's theorem provide an elegant methodology to construct a multivariate distribution, using an arbitrary copula and margins. It provides that using a copula C and margins $F_1, \dots, F_l$ then it is possible to obtain a multivariate distribution function, $F(x) := C(F_1(x_1), \dots, F_l(x_l))$ with margins $F_1, \dots, F_l$.

---

[17] McNeil, Frey & Embrechts (2006) p. 194.

A multivariate distribution built, through a copula and arbitrary margins, is called *meta-distribution*. For example, using a Gaussian copula but arbitrary margins, it is possible to obtain a *meta-Gaussian distribution*. McNeil, Frey & Embrechts (2006) extend the meta-distribution also to the other copulas family

Using the two thousand samples illustrates in the **Figure 11** and normal margins, it is possible to construct four meta distributions. In **Figure 12**, using the quantile function of standard normal distributions, are reported the two thousands simulated points from the four meta distributions: the meta-Gaussian distribution (a), the *meta-Gumbel* distribution (b), the *meta-Clayton* distribution (c) and the *meta-t* distribution (d).

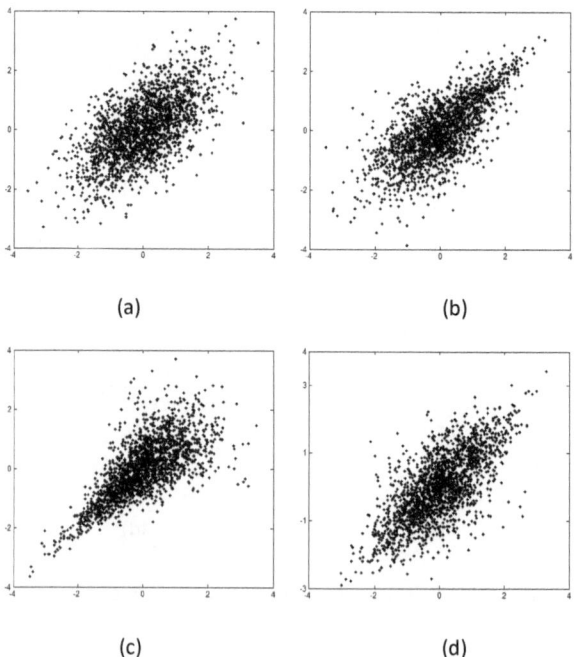

(a)   (b)

(c)   (d)

Figure 12: Two thousand simulated points from four distribution wit normal margins, constructed using the Copula data of figure 11 [18]

---

[18] McNeil, Frey & Embrechts (2006) p. 195.

4.4   Loss Dependence by Means of Copula

Implicitly we have already met copulas in the previous chapter, where we have derived the multivariate distributions in a more familiar and natural way. So the natural question is why are we supposed to work with copula? And are not they superfluous tools?

Copulas are a statistical tools to build up multivariate distributions, which only after long time from their born have started to be applied in credit risk modeling.

Recalling the Sklar theorem (4.2) every multivariate distributions can be represented as copulas; moreover there will be a unique representation, so a unique copula, when the margin distributions are continuous function.

But multivariate distribution do not explicitly separate marginal distributions and correlations as copulas do.

That is why, being the marginal distribution usually known, copula become so popular in the credit risk framework. In particular these are the general framework in static default model.

A static model is a model where default and survival are modeled over a fixed horizon $[0, T]$. Moreover, the set $\{1, ..., I\}$ of obligors is known as well as the individual survival probabilities of the obligors, $(1 - p_i)$. The dependency of default is modeled by the known copula $C(u_1, ... u_I)$.

Given these input it is possible generate different scenario for the credit risk.

From the copula $C(\cdot)$ it is possible to simulate the variables $U_1, ... U_I$ and define the survival of an obligor until time T if:

$$U_1 \leq p_i,$$

where the survival probability of obligor i is defined as:

$$P[U_1 \leq p_i] = p_i.$$

Figure 13 shows the model in a two obligors scenario. The intersection of the survival probabilities divide the unit square into four areas. The pair of random variables (U_1, U_2) drawn from the copula $C(\cdot)$ will represent the coordinates of hypothetical points in the squares. Points falling in the $(S_1 S_2)$ area implies both obligors survive; if they fall in the dark grey area (D_1, D_2) both obligors default; if the points fall in one of the light grey areas (S_1, D_2) or (D_1, S_2) one of the obligors default.

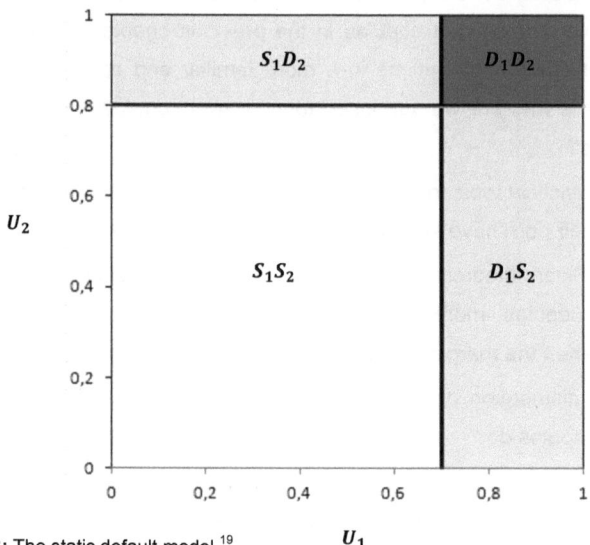

Figure 13: The static default model [19]

It is easy to obtain:

- The probability of all obligors survive is

$$P[U_i \le p_i, \quad \forall i \le I] = C(p_1, \dots, p_I).$$

- The probability that the first k obligors survive is

$$P[U_i \le p_i, \quad \forall i \le k] = C(p_1, \dots, p_k, 1 \dots, 1).$$

- The probability that a subset of obligors $I_S \subset \{1, \dots, I\}$survive is

$$C(u_1, \dots, u_I) \quad where \quad u_i = \begin{cases} p_i & if \ i \in I_S \\ 1 & otherwise \end{cases}$$

------

[19] P. Schonbucher (2003) p. 338.

5 Moment Matching Approximation

5.1 Introduction

In this chapter will be presented the "Analytical Pricing of CDOs in a Multi-factor Setting by a Moment Matching Approach" by Castagna, Mercurio and Mosconi (2012).

The model will be extended and implemented in the next two chapters.

5.2 The Model

In the following the original paper of Castagna et al. is reported.

Consider a portfolio of M obligors, each of them having a unique and distinct loan characterized by an Exposure at Default, EAD_i. The weight of the ith- obligor in the portfolio is therefore: $w_i = EAD_i / \sum_{i=1}^{m} EAD_i$. Each obligor is characterized by a probability of default (PD_i) and a loss given default (LGD_i). In particular the LGD is described by a stochastic random variable Q, assumed to be independent by other sources of riskiness.

The portfolio loss L can be written as the sum of the single obligor's losses:

$$L = \sum_{i=1}^{M} w_i L_i = \sum_{i=1}^{M} w_i Q_i \mathbb{I}_{\{D_i\}} \tag{5.1}$$

where $\mathbb{I}_{\{D_i\}}$ is the indicator function of the default of the ith- obligor.

The stochastic variable Q, following Gordy (2003), is assumed to be distributed as a $Beta(\alpha, \beta)$ where the distinctive parameters have to be chosen such that:

$$\alpha \equiv \mu$$

$$\beta \equiv 1 - \mu$$

where $\mu = E(Q)$, is the mean of the Q's distribution.

According to Gordy (2003), Castagna et al. rewrite the standard deviation σ and the skewness γ, which are functions of α and β in terms of μ. That is:

$$\sigma = \sqrt{\frac{1}{2}\mu(1-\mu)}$$

$$\gamma = \frac{2\sqrt{2}}{3} \cdot \frac{1-2\mu}{\sqrt{\mu(1-\mu)}}.$$

The default D_i is modeled through a structural approach based on the Merton's model (1974) where the obligor's asset dynamics are modeled on one or more common factors, thus obtaining a dependency of the obligor. Castagna *et al.* extend this approach, which is at the base of the international regulation in the Basel II framework, also known as the Asymptotic Single-Risk Factor (ASRF), including *several common factor* and *contagion* effects.

Obligor's default occurs when the random variable X_i, expressing the obligor's asset return, falls below a certain threshold c_i related to the default probability p_i, i.e. $c_i = N^{-1}(p_i)$.

Mathematically speaking, they model the obligor's asset dynamic as a Gaussian random variable:

$$X_i = r_i Y_i + \sqrt{(1-r_i^2)}\xi_i$$

(5.2)

where

$$Y_i = \sum_{k=1}^{N} \alpha_{ik} Z_k \quad \text{and} \quad Z_k \sim N(0,1) \ \ i.i.d.$$

are the common factors describing the effects of several indices (with factor loadings r_i) and represent the systematic part of the asset return, while

$$\xi_i \equiv \xi(\Gamma_i, \epsilon_i) = g_i \Gamma_i + \sqrt{(1-g_i^2)}\epsilon_i$$

is the specific firm effect, which includes a purely idiosyncratic effect $\epsilon_i \sim N(0,1) \ \ i.i.d$, and a contagion effect $g_i \Gamma_i$.

The unconditional correlation of pairs of distinct obligors are:

$$\rho_{ij} = r_i r_j \sum_{k=1}^{N} \alpha_{ik} \alpha_{jk} + \sqrt{(1-r_i^2)}\sqrt{(1-r_j^2)} g_i g_j \sum_{k=1}^{N} \gamma_{ik} \gamma_{jk}.$$

(5.3)

The portfolio loss can be rewritten as:

$$L = \sum_{i=1}^{M} w_i Q_i \mathbb{1}_{\{X_i \leq N^{-1}(p_i)\}}$$

(5.4)

where p_i are the default probabilities of the ith- obligor at time T.

This expression is analytically tractable in order to compute several information about the portfolio of loans. Castagna *et al.*[20] provided an accurately way to estimate the quantile $t_q(L)$ for high confidence level q, through a Taylor Expansion. This method is really accurate for a high level of confidence but it is break down for lower level of confidence such that request for CDO pricing. In this setup lies the Moment Matching method[21] explained in the following.

5.3 The Moment Matching

The Moment Matching (MM) method is useful in working with unknown distributions whose characteristics are difficult to compute. It consists of replacing the original unknown distribution with a proxy distribution, which is easy to work with.

Let X be a random variable, whose distribution is unknown, but assume that it is possible to compute its first n moments $M_1, M_2, ..., M_n$.

Then it is possible to choose a known random variable Y with the same n moments, and replace X by Y. The random variable Y has to be chosen such that it is strictly linked to the unknown random variable X.

The MM approach has several application in the financial field. The most common example is the approximation of the sum of log-normal random variables (whose sum is not log-normal distributed) with a lognormally distributed proxy. This happens in the Black and Scholes framework, where the prices are assumed to be distributed following Log-normal distribution and sum of prices is proxy by a Log-normal distribution. Another example of the common use of the MM approach, is the Asian option pricing.[22]

In the CDO pricing framework Castagna *et al.* use the MM approach to replace the loss portfolio distribution L with a well-known proxy L^*.

In particular, the distribution given by (5.2) is unknown but its moments are analytical computed.

---

[20] Castagna A. ,Mercurio F., Mosconi P. (2009)
[21] Castagna A. ,Mercurio F., Mosconi P. (2012).
[22] E.Levy (1992), S. Turnbull and L.Wakemann (1992).

Then a proxy distribution L^* with known distribution properties is chosen and the MM technique is applied. According to the number n of parameters which define the L^* distribution, the first n moments of L are calibrated on the first n moments of L^*, by solving a system of n equations:

$$\begin{cases} M_1 = M_1^* \\ M_2 = M_2^* \\ \quad \cdot \\ \quad \cdot \\ M_n = M_n^* \end{cases}$$

Solving this system allows to calibrate the n parameters such that the two distributions have the same n moments.

The original distribution given by (5.2) is characterized by the following moments:

First moment M_1

The first moment is simply given by the expected loss:

$$M_1 \equiv E(L) = \sum_{i=1}^{M} w_i \mu_i p_i. \tag{5.5}$$

Second moment M_2

The second moment, starting from its definition, is given by:

$$M_2 \equiv E(L^2) = \sum_{i,j=1}^{M} E(L_i L_j).$$

With some algebra:

$$M_2 = \sum_{i=1}^{M} w_i^2 E(Q_i^2) p_i + \sum_{i \neq j}^{M} w_i w_j \mu_i \mu_j N_2\left(N^{-1}(p_i), N^{-1}(p_j), \rho_{ij}\right) =$$

$$= \sum_{i=1}^{M} w_i^2 \left[E(Q_i^2) p_i - \mu_i^2 N_2(N^{-1}(p_i), N^{-1}(p_i), \rho_{ii})\right] +$$

$$+ \sum_{i=1}^{M} \sum_{j=1}^{M} w_i w_j \mu_i \mu_j N_2\left(N^{-1}(p_i), N^{-1}(p_j), \rho_{ij}\right). \tag{5.6}$$

where $E(Q_i^2) = E(Q_i)^2 + Var(Q_i) = \mu_i^2 + \sigma_i^2$.

Third moment M_3

By definition the third moment is defined as:

$$M_3 \equiv E(L^3) = \sum_{i,j,k=1}^{M} E(L_i L_j L_k).$$

With some algebras:

$$M_3 = \sum_{i=1}^{M} w_i^3 \, E(Q_i^3) p_i + 3 \sum_{i \neq j}^{M} w_i^2 w_j E(Q_i^2) \mu_j \, N_2 \left(N^{-1}(p_i), N^{-1}\left(p_j\right), \rho_{ij} \right) +$$

$$+ \sum_{i \neq j \neq k}^{M} w_i w_j w_k \mu_i \mu_j u_k \, N_3 \left(N^{-1}(p_i), N^{-1}(p_j), N^{-1}(p_k), \Sigma_{3x3} \right). \tag{5.7}$$

where:

$$\Sigma_{3x3} = \begin{pmatrix} 1 & \rho_{ij} & \rho_{ik} \\ & 1 & \rho_{jk} \\ & & 1 \end{pmatrix}.$$

is the symmetric positive definite variance-covariance matrix and

$$E(Q_i^3) = \sigma^3 \gamma + \mu(3\sigma^2 + \mu^2).$$

Fourth moment M_4

By definition the fourth moment is defined as:

$$M_4 \equiv E(L^4) = \sum_{i,j,k,m=1}^{M} E(L_i L_j L_k L_m)$$

With some algebras:

$$M_4 = \sum_{i=1}^{M} w_i^4 E(Q_i^4)p_i +$$

$$+2 \sum_{i \neq j}^{M} [2w_i^3 w_j E(Q_i^3)u_j + 3w_i^2 w_j^2 E(Q_i^2)E(Q_j^2)]N_{23}(N^{-1}(p_i), N^{-1}(p_j), \Sigma_{2x2}) +$$

$$+12 \sum_{i \neq j \neq k}^{M} w_i^2 w_j w_k E(Q_i^2)\mu_j \mu_k N_3(N^{-1}(p_i), N^{-1}(p_j), N^{-1}(p_k), \Sigma_{3x3}) +$$

$$+ \sum_{i \neq j \neq k \neq m}^{M} w_i w_j w_k w_m \mu_i \mu_j \mu_k \mu_m N_4(N^{-1}(p_i), N^{-1}(p_j), N^{-1}(p_k), N^{-1}(p_m)\Sigma_{4x4}). \qquad (5.8)$$

where:

$$E(Q_i^4) = \frac{\mu + 3}{4} E(Q_i^3)$$

and:

$$\Sigma_{4x4} = \begin{pmatrix} 1 & \rho_{ij} & \rho_{ik} & \rho_{im} \\ & 1 & \rho_{jk} & \rho_{jm} \\ & & 1 & \rho_{km} \\ & & & 1 \end{pmatrix}.$$

5.4 Approximating distributions

Castagna *et al.* develop this technique using two proxy distributions, characterized respectively by two and four parameters.

The first one is the Gaussian Copula Large Homogeneous Portfolio approximation (GCLHP) due to Vasicek (1987), which has been already reported in equation (3.32) and it is characterized by two parameters (p, r). The second one is a mixture of two different Vasicek distributions[23].

The original loans portfolio is characterized by n-obligors and by many dependencies. The main idea is to find two parameters p and r such that the first and the second moments of L and L^* are equal, and then replace L by L^*. That is, mapping the original many dependencies into a two parameters distribution $L^*(p, r)$, which entirely codifies the original distribution.

---

[23] In the following only the 2 parameters proxy distribution, which is the one then used in the practical application, is reported. For a complete reading see Castagna *et al.*(2012).

The GCLHP approximation has already been reported in Chapter 3, so as in the (3.32), we derive the CDF for $L^*(p, r)$:

$$F_{p,r}(x) = P[L^* \leq x]$$

$$= N\left[\frac{(N^{-1}[x]\sqrt{1-r}) - N^{-1}[p]}{\sqrt{r}}\right]$$

In order to compute the first n - moments of the approximating distribution $L^*(p, r)$, it is more convenient rewrite the (3.32) as:

$$L^* = N\left[\frac{N^{-1}[p] - \sqrt{r}\,S}{\sqrt{1-r}}\right] \tag{5.9}$$

where $S \sim N(0,1)$.

The n - moments of $L^* \sim F_{p,r}$ are given by:

$$E(L^{*n}) = N_n(N^{-1}(p), \dots, N^{-1}(p), \Sigma_r) \tag{5.10}$$

where $N_n(\dots)$ denotes the n - dimensional normal distributions function and, $\Sigma_r \in R^{nxn}$ is a matrix with 1 on the diagonal and r on all the other positions[24].

By solving the system of two non linear equations, it turns out the calibrated p and r. The original distribution, and so the original loans portfolio, is now replicated by the large portfolio approximation associated with the calibrated parameters.

In the Vasicek model it is possible to compute the expected tranches loss (ETL) by solving analytically the integral in equation (2.4)[25]. Under the large portfolio approximation, the expected loss of a tranche $[L, U]$ is given by:

$$EL_{[L,U]} = \frac{N_2(-N^{-1}(L), N^{-1}(p), \Sigma) - N_2(-N^{-1}(U), N^{-1}(p), \Sigma)}{U - L} \tag{5.11}$$

---

[24] For a complete proof see C.Bluhm, L.Overbeck and C.Wagner (2003) p. 92.
[25] For a complete proof see Anna Schlosser (2010) p. 106.

where N_2 is the bivariate normal distribution function and Σ is the covariance matrix:

$$\Sigma = \begin{pmatrix} 1 & -\sqrt{1-r^2} \\ -\sqrt{1-r^2} & 1 \end{pmatrix}.$$

6 Extensions to the Model

In this Chapter will be presented two extensions I bring to the original model of *Castagna et al.*, representing the main research targets of the thesis.

In the first extension I have rewritten the original model in terms of Archimedean Copulas. The dependencies structure of the original loss distribution has been rewritten in terms of Clayton Copula. The proxy distribution used is the Large Portfolio loss distribution for Archimedean copulas, from which I derived the moments (Proposition 6.1 and 6.2). Finally I derived the ETL formula for this setup (Proposition 6.3).

In the second extension I have provided how the Moment Matching techniques can be useful, in highlighting the sources of risk in the reference portfolio and in managing them to reach the proper risky profile.

6.1 Archimedean Copulas

6.1.1 Moments of the original distribution

In this section I have rewritten the original model in terms of Archimedean Copulas. In particular I provide a Moment Matching techniques for the particular case of Clayton family, which has been chosen for the close formula of the density function of its mixing variable.

The expansion in moments suggests a possible generalization of the moments in terms of Copulas. Let $\mathbb{C}_{i,j,\dots,k;M}$ be a M dimension copula function in which the elements with position $i,j,\dots,k$ are set equal to $p_i, p_j, \dots, p_k$ and the others are set equal to 1. Then, the first (5.5) and the second moment (5.6) can be rewritten as:

$$M_1 \equiv E(L) = \sum_{i=1}^{M} w_i \mu_i \mathbb{C}_{i;M} \tag{6.1}$$

$$M_2 \equiv \sum_{i=1}^{M} w_i^2 E(Q_i^2) \mathbb{C}_{i;M} + \sum_{i \neq j}^{M} w_i w_j \mu_i \mu_j \, \mathbb{C}_{i,j;M} \tag{6.2}$$

In the specific case where the dependencies structure is modeled by a Clayton Copula, as in the extension I provided, by the definition of Clayton family:

$$\mathbb{C}_{i;M} = p_i$$

$$\mathbb{C}_{i,j;M} = \left(p_i^{-\theta} + p_j^{-\theta} - 1\right)^{-\frac{1}{\theta}}.$$

6.1.2 Approximating distribution

In this setup the original distribution is replaced by the Large Portfolio loss distribution for Archimedean copulas due to P. Schonbucher (2002). This is the equivalent of the Vasicek approximation in an Archimedean environment.

For the mathematical derivation of the loss distribution, the complete reading of the article[26] is suggested. In the following, the reader will be assumed confident with the Schonbucher Large portfolio loss distribution approximation.

Let denote $L^S \sim Schonbucher(q, \theta)$ where q is the uniform survival probability and θ is the dependency parameter of the Clayton copula.

According to Schonbucher all obligors have the same conditional survival probability $q(Y)$, defined as:

$$q(Y) = \exp\{-Y\phi(q)\} \qquad (6.3)$$

where Y is the mixing variable. For the LLN the fraction L of default will almost surely be $1 - q(Y)$.

Schonbucher finds that the distribution F and the density f of the Limiting loss distribution are:

$$F_{q,\theta}(x) = P(L^S \leq x) = G\left(-\frac{\ln(1-x)}{\phi(q)}\right), \qquad (6.4)$$

$$f_{q,\theta}(x) = \frac{1}{(1-x)\phi(q)} g\left(-\frac{\ln(1-x)}{\phi(q)}\right), \qquad (6.5)$$

where G and g are the distribution and the density function of the mixing variable Y. In the Clayton case G is a Gamma distribution with parameters $\left(\frac{1}{\theta}, 1\right)$.

---

[26] P. Schonbucher (2002).

Proposition 6.1 (first moment) *The first moment of the Large portfolio loss distributions for Archimedean copulas, in the Clayton case is:*

$$M_1^S = 1 - \frac{1}{(1 + \phi(q))^{\frac{1}{\theta}}}. \tag{6.6}$$

Proof.

$$E[L] = E[1 - q(Y)] = 1 - E[q(Y)] = 1 - \int_{-\infty}^{\infty} q(Y)g(y)dy$$

$$= 1 - \int_{-\infty}^{\infty} e^{-y\phi(q)} \frac{1}{\Gamma(1/\theta)} e^{-y} y^{(1-\theta)/\theta} dy = 1 - \int_{-\infty}^{\infty} \frac{1}{\Gamma(1/\theta)} e^{-y(1+\phi(q))} y^{(1-\theta)/\theta} dy$$

Multiplying and dividing the second term by $(1 + \phi(q))^{\frac{1}{\theta}}$, it is possible to obtain a density function of a $Gamma \sim \left(\frac{1}{\theta}, \frac{1}{1+\phi(q)}\right)$, whose integral over $\mathbb{R}$ is 1.

$$E[L] = 1 - \frac{1}{(1+\phi(q))^{\frac{1}{\theta}}} \int_{-\infty}^{\infty} (1+\phi(q))^{\frac{1}{\theta}} \frac{1}{\Gamma(1/\theta)} e^{-y(1+\phi(q))} y^{(1-\theta)/\theta} dy$$

$$= 1 - \frac{1}{(1+\phi(q))^{\frac{1}{\theta}}}.$$

Proposition 6.2 (second moment) *The second moment of the Large portfolio loss distributions for Archimedean copulas, in the Clayton case is:*

$$M_2^S = 1 - \frac{2}{(1+\phi(q))^{\frac{1}{\theta}}} + \frac{1}{(1+2\phi(q))^{\frac{1}{\theta}}}. \tag{6.7}$$

Proof.

$$E[L^2] = E\left[(1 - q(Y))^2\right] = E\left[1 - 2q(Y) + (q(Y))^2\right] = 1 - 2E[q(y)] + E[q(y)^2]$$

$$= 1 - \frac{2}{(1+\phi(q))^{\frac{1}{\theta}}} + \int_{-\infty}^{\infty} \left(e^{-y\phi(q)}\right)^2 g(y)dy$$

$$= 1 - \frac{2}{(1+\phi(q))^{\frac{1}{\theta}}} + \int_{-\infty}^{\infty} e^{-2y\phi(q)} \frac{1}{\Gamma(1/\theta)} e^{-y} y^{(1-\theta)/\theta} dy$$

$$= 1 - \frac{2}{(1+\phi(q))^{\frac{1}{\theta}}} + \int_{-\infty}^{\infty} \frac{1}{\Gamma(1/\theta)} e^{-y(1+2\phi(q))} y^{(1-\theta)/\theta} \, dy$$

Multiplying and dividing the third term by $(1 + 2\phi(q))^{\frac{1}{\theta}}$, it is possible to obtain a density function of a Gamma $\left(\frac{1}{\theta}, \frac{1}{1+2\phi(q)}\right)$, whose integral over $\mathbb{R}$ is 1.

$$E[L^2] = 1 - \frac{2}{(1+\phi(q))^{\frac{1}{\theta}}} + \frac{1}{(1+2\phi(q))^{\frac{1}{\theta}}} \int_{-\infty}^{\infty} (1+2\phi(q))^{\frac{1}{\theta}} \frac{1}{\Gamma(1/\theta)} e^{-y(1+2\phi(q))} y^{(1-\theta)/\theta} \, dy$$

$$= 1 - \frac{2}{(1+\phi(q))^{\frac{1}{\theta}}} + \frac{1}{(1+2\phi(q))^{\frac{1}{\theta}}}.$$

Finally remembering that the generator of a Clayton copula is $\phi(t) = t^{-\theta} - 1$, the first (6.6) and the second moments (6.7) became:

$$M_1^S = 1 - q \tag{6.8}$$

$$M_2^S = 1 - 2q + \frac{1}{(2q^{-\theta} - 1)^{\frac{1}{\theta}}}. \tag{6.9}$$

In the Schonbucher Model it is possible computing analytically the integral in (2.4).

Proposition 6.3 *In the Large portfolio loss distributions for Archimedean copulas, the Expected Tranches Loss for the Clayton case assuming zero recovery rate, is given by:*

$$E[(L^S - K)^+] = (1 - K)\left[1 - G_1\left(-\frac{\ln(1-k)}{\phi(q)}\right)\right] - \frac{1}{[1+\phi(q)]^{-\frac{1}{\theta}}}\left[1 - G_2\left(-\frac{\ln(1-k)}{\phi(q)}\right)\right], \tag{6.10}$$

where:

- $G_1 \sim Gamma\left(\frac{1}{\theta}; 1\right)$
- $G_2 \sim Gamma\left(\frac{1}{\theta}; \frac{1}{[1+\phi(q)]}\right).$

Proof.

$$E[(L^S - K)^+] = E[(1 - q(Y) - K)^+] = E[(1 - e^{-y\phi(q)} - K)^+]$$

$$= \int_{-\frac{ln(1-k)}{\phi(q)}}^{+\infty} [1 - e^{-y\phi(q)} - K]g(y)dy = \int_{-\frac{ln(1-k)}{\phi(q)}}^{+\infty} [1 - K]g(y)dy - \int_{-\frac{ln(1-k)}{\phi(q)}}^{+\infty} e^{-y\phi(q)} g(y)dy$$

Multiplying and dividing the second term by $[1 + \phi(q)]^{-\frac{1}{\theta}}$ it is possible to obtain the density function of a $Gamma\left(\frac{1}{\theta} ; \frac{1}{1+\phi(q)}\right)$.

$$= (1 - K) \int_{-\frac{ln(1-k)}{\phi(q)}}^{+\infty} g(y)dy - \frac{1}{[1 + \phi(q)]^{-\frac{1}{\theta}}} \int_{-\frac{ln(1-k)}{\phi(q)}}^{+\infty} \frac{[1 + \phi(q)]^{-\frac{1}{\theta}}}{\Gamma(1/\theta)} e^{-y[1+\phi(q)]} y^{(1-\theta)/\theta} dy$$

$$= (1 - K)\left[1 - G_1\left(-\frac{ln(1 - k)}{\phi(q)}\right)\right] - \frac{1}{[1 + \phi(q)]^{-\frac{1}{\theta}}}\left[1 - G_2\left(-\frac{ln(1 - k)}{\phi(q)}\right)\right].$$

6.2 Risk Management

In this further extension of the original paper I provide a new reading key of the MM approach in terms of risk covering. In particular, through a reverse MM procedure, I provide a method to modify sources of risk in order to reach the desirable risk profile.

6.2.1 Risk Measures

The Economic Capital is the amount of capital to be set apart and to be immediately cashable to absorb losses. This can be determined by different measures of risk. One of these is the VaR.

VaR is strictly linked to a probability distribution of losses over a given period and to the statistical concept of quantile.

Definition

Let X *be a r.v. and* $F_X(x) = P(X \leq x)$ its distribution function. If F is invertible (i.e., continuous and without jumps) the quantile of X of order α is:

$$q_\alpha(X) = \inf\{x : P(X \leq x) \geq \alpha\}.$$

In an economic framework where $F_X(x)$ is the probability distribution of losses, the **Value-at-risk** of order α is defined as:

$$VaR_\alpha(X) = -q_\alpha(X).$$

The minus sign translates (negative) losses in (positive) risks

6.2.2 The method

The Moment Matching techniques has been used until now in a pricing CDO framework, but it is a general tool to replace whatever distributions with an approximating one, which is usually chosen for its better known properties.

For example assume that a financial institution has a loans portfolio and it has to evaluate the Economic Capital (EC) required to face unexpected losses. Through a MM approach, the original loss distributions could be replaced with an approximating one, whose better known properties make the computation of the risk measures easier, i.e. the VaR.

Let the portfolio of a financial institution be constituted by M obligors, but differently from before the aim is no more the securitization activity, but the evaluation of the EC required to face unexpected losses.

The loss distribution of a similar portfolio is as before described by the equation (5.4) and as before, it is replaced by a proxy one, for example the Large Portfolio Approximation distribution $L^*(p, r)$, whose parameters p and r are calibrated by a MM approach.

Replacing the original distributions L, with the proxy $L^*(p, r)$, makes the computation of the measures associated to the unexpected loss risk, easier.

For any given level of confidence α, the quantile of order α of the random variable $L^* \sim F_{p,r}(x)$ is given by[27]:

---

[27] Bluhm, Overbeck and Wagner (2003) p. 90.

$$q_\alpha(L^*) = N\left(\frac{N^{-1}(p) + \sqrt{r}q_\alpha(Y)}{\sqrt{1-r}}\right)$$

(6.11)

where $Y \sim N(0,1)$ and $q_\alpha(Y)$ is simply the quantile of order α of the standard normal distribution.

Suppose a financial institution has a portfolio of loans, whose loss distribution can be written as in equation (5.4) and suppose to obtain by a MM approach, the calibrated parameters $p = 0.05\%$ and $r = 10\%$. The 99.98% quantile can be easily computed by (6.11) as $q_{99.98\%}(L^*) = 0.3832$.

Let us assume the Economic Capital is considered too huge for such financial institution, then it can fix the adequate EC and then by a reverse MM procedure, modifies the composition of the portfolio of loans, to reach its adequate risk profile.

In a similar contest the MM approach could be really useful for the financial institution because the moment equations highlight the relevance of several sources of risks.

The equation of the second moments

$$M_2 = \sum_{i=1}^{M} w_i^2\, E(Q_i^2)p_i + \sum_{i \neq j} w_i w_j \mu_i \mu_j\, N_2\big(N^{-1}(p_i), N^{-1}(p_j), \rho_{ij}\big)$$

highlights several sources of risk:

i.   the concentration risk (the weights);

ii.   the recovery risk (the first moment of the stochastic variable Q);

iii.   the default risk (the individual default probabilities)

iv.   the correlation risk (joint probabilities of default of all possible combinations of pairs of obligors).

The main idea is performing the MM approach but in a reverse procedure. In the CDO pricing setup, starting from the original moments, the proxy are found, via a calibration method.

Established the maximum Economic Capital which a financial institution is available of to set aside for a given portfolio of loans, by a reverse MM procedure it is possible to find the optimal value of the original moments, which should imply such EC. Then, the optimal value of the original moments will be obtained, by modifying the composition of the original portfolio, i.e. by modifying the sources of risk highlighted by Moments expansion.

61

In the previous example, $q_{99.98\%}(L^*) = 0.3832$ leads to a $EC_{99.98\%} = 383200 \ Euro$ for a total amount of the loans portfolio equal to 1000000 Euro.

Assuming that, for such confidence level, the institution is available to set aside just 300000 Euro, this means that the optimal quantile should be $q_{99.98\%}^*(L^*) = 0.30$.

By equation (5.2), fixing $q_{99.98\%}^*(L^*) = 0.30$ and leaving the uniform default probability equal to 5%, it is possible to obtain the optimal uniform correlation $r^* = 0.105203$.

So the optimal parameters to obtain the desired quantile, and thus the supportable Economic Capital, is $(p^*, r^*) = (0.05 \ ; \ 0.1052)$.

These parameters imply that the first and the second moments of the approximating distribution are $M_1^* = 0.05$ and $M_2^* = 0.003778565$.

The moments of the original institution have to be equated to the moments of the proxy distribution by a reverse MM procedure.

$$
\begin{cases}
M_1^* = \displaystyle\sum_{i=1}^{M} w_i \, \mu_i p_i \\
M_2^* = \displaystyle\sum_{i=1}^{M} w_i^2 \, E(Q_i^2) p_i + \sum_{i \neq j}^{M} w_i w_j \mu_i \mu_j \, N_2\big(N^{-1}(p_i), N^{-1}(p_j), \rho_{ij}\big) .
\end{cases}
$$

The moment expansion allows to modify the risk sources in order to obtain the target moments.

In the previous numerical example, the target moments are respectively $M_1^* = 0.05$ and $M_2^* = 0.003778565$. The financial institution should modify the composition of its loans portfolio in order to reach this level. That is, by reducing the weights of obligors the company will reduce the concentration risk as well as it will reduce the default risk by reducing the weight of obligors with higher probability of default.

7 Implementation

In the following chapter a numerical implementation of the Moment Matching techniques will be proposed, aiming to price the tranches of a CDX.

In the first section the pricing of the CDX tranches will be obtained by implementing the original model of Castagna *et al.* reported in chapter 5. This is the first time that the method is numerically implemented with real data. The first implementation will be denoted in the following as Gaussian, referring to the dependence chosen.

In the second section the pricing of the CDX will be obtained by implementing the new model I derived in chapter 6. Being this model, the extension of the original one in terms of Clayton copula, will be denoted in the following as Clayton.

7.1 Gaussian implementation

7.1.1 Data

The MM techniques is now applied to price the tranches of a CDX. Appendix A reports the 125 names and the quotes of the CDS contracts with maturities up to 10 years. The contracts are expressed in basis points and the quotes were collected on July 3rd 2007.

For each names the default probabilities up to 10 years are bootstrapped and collected in the Appendix B.

Appendix C reports the discount factors.

These data were already available in the University of Bologna and so were directly used as starting point for all the subsequent computations.

7.1.2 Bootstrapping the default term structure

A CDS is a credit derivative instrument in which one party called "protection buyer" pays a premium, usually on a running base, to another party called "protection seller" to obtain protection if a credit event, i.e. default, occurs. The CDS, as a typical credit derivative contract, is characterized by two streams of payments. The stream of payment due by "protection buyer" is called the "premium leg" and the stream of payment due by the other party the "premium seller" is called the

63

"protection leg". Roughly speaking, the protection buyer pays a premium on a running base until a default event occurs. The protection seller receives the premium and has to correspond an amount to the protection buyer in the case a default occurs.

Let $Q(t)$ be the survival probability of the issuer up to time t, we can define the present value of the two legs. The present value of each payment of the protection leg is:

$$B(t, t_i)[Q(t_{i-1}) - Q(t_i)]LGD$$

where $Q(t_{i-1}) - Q(t_i)$ is the probability of observing default between time t_{i-1} and t_i , LGD is the payment that has to be done and $B(t, t_i)$ is the discount factor. By the same, the present value of each payment in the premium leg is given by:

$$B(t, t_i)Q(t_{i-1})s$$

where s is the premium and $Q(t_{i-1})$ is the probability to survive until that period. As it is usual in credit derivative contracts, the fair premium is the one balancing the present value of the protection leg and the premium leg, that is:

$$\sum_{i=1}^{n} B(t, t_i)Q(t_{i-1})s_n = \sum_{i=1}^{n} B(t, t_i)[Q(t_{i-1}) - Q(t_i)]LGD \qquad (7.1)$$

where s_n is the premium charged for n periods of protection.

The CDS market competes with equity in terms of the name's information provided, because the underlying of the contract is the whole debt of the firm. In addition, CDS market is a liquid market at least for the most important firms, and looking at the CDS quotes, allows to determine how the market quotes the default probability term structure of the name. That is the bootstrapping procedure which is used to recover the default probability of the names included in the CDX.

In particular, assuming that a CDS spread s_1 is quoted for protection over a one-year horizon[28], by the previous definition of a CDS spread we have:

$$B(t, t_1)s_1 = B(t, t_1)[1 - Q(t_1)]LGD$$

---

[28] Assume for simplicity that the spread is paid in one instance at the end of the year and it is paid even if default occurs.

from which the survival probability is recovered as:

$$Q(t_1) = 1 - \frac{S_1}{LGD}.$$

The survival probability up to one year can be now substituted in the two-years CDS

$$\sum_{i=1}^{2} B(t, t_i)Q(t_{i-1})S_n = \sum_{i=1}^{n} B(t, t_i)[Q(t_{i-1}) - Q(t_i)]LGD$$

in order to obtain the survival probability up to two years $Q(t_2)$.

In general

$$Q(t_n) = Q(t_{n-1})\left[1 - \frac{S_n}{LGD}\right] - \frac{S_n - S_{n-1}}{B(t, t_n)}\sum_{i=1}^{n-1} B(t, t_i)Q(t_{i-1}) \qquad (7.2)$$

Bootstrapping consists in extracting the default probability by just applying the spread formula using the spread quoted in the market. This is the procedure used to obtain the default probability term structure of the 125 names included in the CDX. Table 2 in the Appendix shows the results obtained[29].

7.1.3 Regressions

The asset returns dynamics, according to the most common literature are modeled as a Gaussian random variable where a *contagion* term is included. In the CDX pricing several simplifications are carried to the original model, in order to make the implementation easier.

Asset value process is completely described by only one single factor common to all counterparties. In addition the *contagion* terms is now assumed to be worthless. The asset value process (5.2) is now simplified as:

$$X_i = r_i Y + \sqrt{(1 - r_i^2)}\epsilon_i \qquad (7.3)$$

where the single factor is denoted by $Y \sim N(0,1)$ and X_i are the standardized asset value log-returns.

---

[29] The default probabilities are simply $[1 - Q(t_n)]$.

Equation (7.3) is nothing but a standard linear regressions and for each name the historical standardized asset log-returns are regressed on the standardized factor log-returns.

Because the CDX is an American index which includes only North America "names", the single factor chosen is the S&P500 index. The regressions are computed using 10 years of historical series. More precisely, the data are collected for the period 3/04/1997 - 3/07/2007 using the data provider "Datastream".

Unfortunately the historical data are not available for all the 125 names, but only for 88 of them. This is due to the fact that some CDS are referred to unlisted branch of multinationals. In that case it has been preferred excluding the names from the CDX . Of course, this makes the computation of the tranches prices less accurately, because not all "firms" are included, but the main idea of how to implement the method still lasts. Table 4 in the Appendix reports the 88 coefficients r_i.

By just applying to the simplified process in (7.3), the definition of correlation

$$\rho_{ij} = corr(i,j) = \frac{E[X_i - \mu_{X_i}]E[X_j - \mu_{X_j}]}{\sigma_{X_i}\sigma_{X_j}}$$

the equation in (5.3) simply becomes:

$$\rho_{ij} = r_i r_j.$$

$$(7.4)$$

The large number of obligors leads to difficulties in the numerical computation of the moments. In particular the difficulties lie in the second moment. By definition moments (5.6) the computation of the second moment requires, for 88 obligors:

- the computations of 3828 pairs of asset correlations ρ_{ij}.
- the consequent computation of 3828 bivariate joint distributions $N_2(N^{-1}(p_i), N^{-1}(p_j), \rho_{ij})$.

For this reason the model has been further simplified by dividing the 88 "names" into homogeneous classes, where firms belonging to the same class are characterized by the same asset dynamic. This reduces the equation (7.3) to the following form in terms of classes:

$$X_I = r_I Y + \sqrt{(1 - r_I^2)}\epsilon_I$$

(7.5)

where the index $I = A, B, \ldots, M$ represent the class.

The obligors are classified into different risky assets according to several criteria.

The setup of the equation (7.5) looks like a linear regression again. The only difference is that the class value X_I cannot be observed. In economic literature, this problem is known as *hidden variables regression* but here it is just assumed that the coefficient regression r_I, the request term for computing the correlation, is given by the arithmetic mean of the coefficient r_i of the firms belonging to the I class.

The weight of each class is given by $w_I = I/M$, where I is the number of obligors grouped in the same class and M is the total number of obligors.

Each class is then characterized by the same probability of defaults p_A given by the arithmetic mean of the default probabilities of the class elements'. In the classes framework, the correlation between classes is given as in (7.4) by:

$$\rho_{AB} = r_A r_B$$

(7.6)

but using classes the correlation matrix is hugely reduced. By assuming 4 classes, the pairs correlations are reduced to 6, as well as the bivariate joint distributions.

7.1.4 Clustering

In the tranches pricing the classes are formed in three different ways: Sectors, *K-means*, and *Neural network*. The first one is simply a qualitative method, where firms are grouped according to the sector they belong to. The other two methods are mathematical ones of grouping data.

Sectors

Firstly, via a qualitative criterion, the obligors are classified into different risk classes according to the "name" sector. Between the same sector class no distinction is made between the obligors.

The 88 firms are grouped in 4 sector classes: *financial & utilities*[30], *non-cyclical*, *cyclical*, *industrial*.

Table 2 reports the main information of the classes. For each class a default probabilities term structure has been recovered by averaging the single term structures of the class elements.

SECTORS							
CLASS	Firms	Weight	r_1	p^3	p^5	p^7	p^{10}
financial & utilities	21	0.24	0.42	0.01	0.02	0.04	0.08
non cyclical	17	0.19	0.33	0.01	0.03	0.07	0.13
Cyclical	18	0.20	0.46	0.02	0.05	0.10	0.17
industrial	32	0.36	0.44	0.01	0.03	0.06	0.12

Table 2: Sector: classes information

Table 3 reports the correlation matrix between the classes. The correlation elements, computed via the equation (7.6), are necessary for the computation of the two dimensional normal distributions requested in the second moment.

	financial & utilities	non cyclical	cyclical	industrial
financial & utilities	1	0.14	0.20	0.19
non cyclical	0.14	1	0.15	0.15
cyclical	0.20	0.15	1	0.15
industrial	0.19	0.15	0.21	1

Table 3: Sector: correlation matrix

K-Means

K-means clustering is a partitioning method which groups n observations into k clusters. The function K-means partitions a vector (or a matrix) of n observations into k mutually exclusive clusters and returns a vector of indices, relating to the cluster assigned to each observation.

K-means algorithm treats each element of the observation's vector as an object with a location in space. The main idea of the K-means function is to find a partitioning such that points within the same group are as close as possible to

---

[30] The latter class has formed by join financial and utilities firms together. This is due to the minimum number of elements a class need to be considered large.

each other, and as far as possible from elements of other groups. Each cluster is characterized by a particular point called *centroid*, i.e. the point which minimizes the sum of distances of each points and the *centroid* itself.

In general, given a set of points $x_1, x_2, \dots, x_n \in R^d$ and an integer k (with $k < n$) the k-means function returns a partitions of the points into $S_1, \dots, S_k$ such that

$$\sum_{j=1}^{k} \sum_{x_i \in S_j} \|x_i - \mu_j\|^2 \tag{7.7}$$

is minimal[31]. Notice that μ_j is the mean of the points belonging to the cluster S_j.

In the application the K-means function is applied to the set of points $x_1, x_2, \dots, x_n \in R^2$ where $x_i = (r_i, p_i^1)$. k is arbitrary chosen equal to 3.

The matrix $\mathbb{M}^{(88,2)}$ is partitioned by the algorithm into 3 classes A, B, C containing respectively [32 19 37] elements.

Table 4 reports the main information of the classes.

K-MEANS							
CLASS	Firms	w_A	r_A	p^3	p^5	p^7	p^{10}
A	32	0.36	0.30	0.01	0.03	0.06	0.11
B	19	0.22	0.56	0.01	0.02	0.05	0.09
C	37	0.42	0.45	0.02	0.04	0.08	0.15

Table 4: K-means: classes information

Table 5 reports the correlation matrix between the classes.

	A	B	C
A	1	0.17	0.14
B	0.17	1	0.26
C	0.14	0.26	1

Table 5: K-means: correlation matrix

Neural Networks

Self-Organizing Map (SOM) is one of the most important artificial neural networks (ANN) and it was first developed by T.Kohonen in 1982.

---

[31] $\|\dots\|$ is the Euclidian distance.

SOM is trained using *unsupervised training* techniques, in which the networks form the classification of the data without external help. It is commonly assumed that the class are defined by the input patterns, which shared common features, and that the network is able to identify those common features between inputs.

The main target of SOM is transforming an incoming input matrix of arbitrary dimensions into a low-dimensional (typically one or two-dimensional) discrete representation of the input space of the training samples, called a *map*.

SOM, differently from other type of ANN, preserve the topological properties of the input space. A SOM consists of components called *nodes* (or *neurons*) and associated to each node a weight vector of the same dimension as the input data vectors and a position in the map space. The SOM maps an high dimensional input space in a low dimensional map space. The procedure for placing a vector from data space onto the map is to first find the node with the closest weight vector to the vector taken from data space. Once the closest node is located it is assigned the values from the vector taken from the data space.

As in the k-means application the input matrix is $\mathbb{M}^{(88,2)}$, where the first column is represented by the 88 regression coefficient r_i and the second column is formed by the 88 probability of default within one year $88\,p_i^1$. The result is a two dimensional array corresponding to the four clusters A, B, C, D containing respectively [34 13 18 23] elements.

Table 6 reports the clusters information.

NEURAL NETWORKS							
CLASS	Firms	w_A	r_A	p^3	p^5	p^7	p^{10}
A	23	0.26	0.35	0.01	0.03	0.06	0.11
B	18	0.20	0.57	0.01	0.03	0.05	0.09
C	13	0.15	0.25	0.01	0.03	0.06	0.11
D	34	0.39	0.46	0.02	0.04	0.08	0.15

Table 6: NN: classes information

Table 7 reports the correlation matrix of the classes.

	A	B	C	D
A	1	0.20	0.09	0.16
B	0.20	1	0.14	0.26
C	0.09	0.14	1	0.11
D	0.16	0.26	0.11	1

Table 7: NN: correlation matrix

7.1.5 Original Moments

The original distribution of the portfolio loss is given by equation (5.4) and the first and the second moments, necessary for the calibration, are respectively derived in equations (5.5) and (5.6).

In addition to the simplification on the asset dynamics summarized in (7.2), further assumptions on the original distribution have been done to implement the model.

The stochastic variable Q associated to the LGD is assumed to be a deterministic variable. In addition $LGD = 0.6$ for all obligors.

The first and the second moment are rewritten in the light of the model simplification and are written in terms of class $\{I = A, \ldots, M\}$. That is:

$$M_1 = \mu \sum_{I=1}^{M} w_I p_I \tag{7.8}$$

$$M_2 = \mu^2 \sum_{I=1}^{M} w_I^2 p_I + \mu^2 \sum_{I \neq J}^{M} w_I w_J N_2 \left(N^{-1}(p_I), N^{-1}(p_J), \rho_{IJ} \right). \tag{7.9}$$

Clustering and so reducing the problem dimension, reduces the complexity in the numerical computation.

The advantages of clustering are really clear especially regarding the second moment computation. Before clustering the correlation between 88 obligors were 3828 which implies that, this is also the number of $N_2(\ldots)$ which were supposed to compute.

Grouping the 88 firms into 4 clusters allows to reduce the number of $N_2(...)$ from 3828 to just 6.

The moments (7.8) and (7.9) are computed for all the three methods of clustering and for each methodology the moments are computed for the different default probabilities p^3, p^5, p^7 and p^{10}.

Table 8 reports the first and the second moments of the original distributions for different clustering methods.

p^l	SECTORS		K-MEANS		NEURAL NETWORKS	
	M1	**M2**	**M1**	**M2**	**M1**	**M2**
3	0.007069	0.001210	0.007069	0.00166	0.007069	0.001366
5	0.020726	0.003921	0.020726	0.005173	0.020726	0.004326
7	0.040329	0.008472	0.040329	0.010724	0.040329	0.009115
10	0.073329	0.017577	0.073329	0.021288	0.073329	0.018619

Table 8: Original moments for different clustering methods

7.1.6 Moment Matching

Once computed the original moments, they are calibrated on the moments of the approximating distributions, via a MM technique.

The distribution chosen as proxy of the original one, is the Vasicek's Large Portfolio Approximation (3.32). That is, a distribution characterized by two parameter: the uniform probability of default p and the uniform correlation of default r. By applying the equation (5.10) the first and second moments of the proxy distribution $L^*(p, r)$ are given by:

$$M_1^* = N(N^{-1}(p)) \tag{7.10}$$

$$M_2^* = N_2(N^{-1}(p), N^{-1}(p), r). \tag{7.11}$$

Solving the non linear system $M = M^*$, where M and M^* are the two-dimension vectors of the moments of the two distribution L and L^*, it is possible to obtain the calibrated p and r.

Table 9 reports the calibration results for the different clustering methods and for different probabilities of default.

p^i	SECTORS		K-MEANS		NEURAL NETWORKS	
	p	r	p	r	p	r
3	0.0071	0.0012	0.0071	0.0017	0.0071	0.0014
5	0.0207	0.0039	0.0207	0.0052	0.0207	0.0043
7	0.0403	0.0085	0.0403	0.0107	0.0403	0.0091
10	0.0733	0.0176	0.0733	0.0213	0.0733	0.0186

Table 9: Calibration results

7.1.7 Pricing CDX Tranches

Once the parameters of the approximating distributions are calibrated, it is possible to derive the final tranches price. As it is common for the CDX the tranches are divided into $0 - 3\%, 3 - 7\%, 7 - 10\%, 10 - 15\%, 15 - 30\%$ and $30 - 100\%$.

The tranches spread is computed as in equation (2.7), where the terms $EL_{[L,U]}(t_i)$ are called Expected Tranches Loss (ETL) and are defined as:

$$EL_{[L,U]}(t_i) = \frac{E\left[(L_{t_i} - L)^+\right] - E\left[(L_{t_i} - U)^+\right]}{B - A} \qquad (7.12)$$

In order to simplify the computation, it has been assumed that the premia are paid in a unique date which corresponds to the maturity date.[32]

In the framework of the Large Portfolio Approximation, the expected losses are given by:

$$E\left[(L_{t_i} - x_0)^+\right] = N_2(N^{-1}(p), -N^{-1}(x_0), -\sqrt{1 - r^2}) \qquad (7.13)$$

and the ETL in the GCLHPA is already reported in equation (5.11).

Table 10 reports the ETL, using the Vasicek distribution as $L^*(p, r)$, for the different clustering methods.

---

[32] It is common practice that the premia are quarterly paid for each year until the maturities.

ETL SECTOR					
years	0-3%	3-7%	7-10%	10-15%	15-30%	30-100%
3	0.2305	4.99E-05	1.31E-06	1.50E-07	1.63E-09	2.72E-14
5	0.4493	0.0016	0.0004	0.0003	4.46E-05	3.49E-07
7	0.6740	0.0037	0.0013	0.0010	0.0002	2.078E-06
10	0.8845	0.0073	0.0031	0.0028	0.0006	7.364E-06

ETL K-MEANS					
years	0-3%	3-7%	7-10%	10-15%	15-30%	30-100%
3	0.1268	0.0360	0.0183	0.0105	3.78E-03	2.57E-04
5	0.3735	0.1202	0.0573	0.0299	0.0085	3.19E-04
7	0.5745	0.2543	0.1398	0.0801	0.0261	1.17E-03
10	0.8144	0.4883	0.3022	0.1841	0.0621	2.54E-03

ETL NEURAL NETWORKS					
years	0-3%	3-7%	7-10%	10-15%	15-30%	30-100%
3	0.1268	0.0360	0.0183	0.0105	3.78E-03	2.57E-04
5	0.4254	0.1188	0.0477	0.0212	4.41E-03	8.48E-05
7	0.6533	0.2735	0.1316	0.0652	0.0156	3.48E-04
10	0.8643	0.5285	0.3115	0.1754	0.0486	1.22E-03

Table 10: Expected Tranche Loss for different clustering methods

These ETL term structures, by (2.7) lead to the final tranches spread. For the equity tranche it is common practice ,fixing the periodic spread to 500 basis point and then quote the upfront needed to make the contract fair.

The tranches spread are collected in **Table 11**.

ETL SECTOR					
0-3%	3-7%	7-10%	10-15%	15-30%	30-100%
5%	0.063553%	0.031523%	0.015861%	0.003744%	7.08E-05%
ETL K-MEANS					
0-3%	3-7%	7-10%	10-15%	15-30%	30-100%
5%	0.056966%	0.031225%	0.017877%	0.00571%	0.000231%
ETL NEURAL NETWORKS					
0-3%	3-7%	7-10%	10-15%	15-30%	30-100%
5%	0.062475%	0.0318%	0.016666%	0.004357%	0.000108%

Table 11: Tranches spread for different clustering methods

Following the practice of fixing the spread of the equity tranche at the value of 500 basis points, the upfronts charged on the equity tranches are $U^{Sector} = 0.6656, U = 0.5978$ and $U^{NN} = 0.6410$.

7.2 Clayton Approach

In this section the pricing of the CDX tranches will be obtained implementing the methods I derived in chapter 6.

Differently, from the Gaussian implementation, in the following the tranches spread will be obtained only for the K-means clustering. The choosing of this particular clustering over the three previous classifications is due to the smallest number of classes it provides.

The moments of the original distributions are computed as in (6.1) and (6.2).

For the particular case of Clayton copula, it is possible to rewrite the two-dimensional copula in (6.2) as:

$$\mathbb{C}_{i,j;M} = \left(p_i^{-\theta_{ij}} + p_j^{-\theta_{ij}} - 1 \right)^{-\frac{1}{\theta_{ij}}},$$

where each pairs of obligors are characterized by the Clayton parameter θ_{ij}.

In the implementation the parameters θ_{ij} will be recovered for each pair of obligors, using the methods developed by Frees and Valdez (1998) in which an Archimedean copula is fitted to a bivariate series of data.

The method consists of two steps. In the first step, among all the Archimedean copulas, the best copula fitting data is chosen. In the second step, the parameters of the best fitting copula are found via Maximum Likelihood Estimation (MLE).

Being the implementation based on a Clayton copula, the first step has been removed, and then the parameter θ, the unique parameter defining a Clayton copula, is recovered by MLE.

The copula has to be fitted to bivariate data, which in this setup are represented by CDS time series. The dependencies between each pair of obligors is represented by the ML estimated of θ_{ij}, which is the parameter of the Clayton copula fitting the bivariate time series CDS.

As in the Gaussian implementation, the large number of the reference portfolio, leads to several difficulties. This means, that for the 88 obligors should be collected 3828 pairs of CDS time series and then 3828 Clayton copula should be fitted to them, leading to the MLE of the 3828 parameters θ_{ij}.

In addition this step should be repeated for the different maturities of the CDS. In our case, CDS for 3,5,7 and 10 years maturities are requested according to the maturities of the CDX.

7.2.1 Construction of the Data

As in the Gaussian setup the problems of the large portfolio has been solved by dividing obligors into homogeneous classes. Differently from the Gaussian setup, in the following the implementation has been conducted only for the *K-means* clustering. The choosing of this particular division is due to the smallest number of group it provided as well as to the bigger number of elements grouped in each class.

The time series of the 88 5 year CDS are collected by Bloomberg. More precisely, the time series cover the period from 1/03/2004 to 01/07/2007. The latter is the date from which the individual default probability are bootstrapped.

Unfortunately, the time series of the CDS with maturities 3, 7 and 10 years were not suitable for an implementation. So in the following the dependencies of time series are assumed to be equal indifferently by the maturity of the CDS.

The characteristics of three classes provided by *K-means* clustering are the same reported in **Table 4.**

By clustering the total amount of obligors into 3 classes, the estimation of the parameters θ_{IJ}[33] is reduced. In particular, decreasing from 3828 to just 3.

The Clayton copula has to be fitted to the bivariate data, which are now represented by the CDS time series of the centroid. The centroid is a statistical characteristics of each class and its time series has been artificially constructed, as arithmetic mean of the CDS time series of the firms belonging to the same class.

The bivariate date have been fitted by Clayton copula, whose parameters are reported in **Table 12.** The estimation of the Clayton parameters has been

---

[33] Notice that the parameter is now referred to the classes I, J.

implemented using an Excel program, based on the work of Frees and Valdez, developed by the Prof. U. Cherubini of University of Bologna.

θ_{IJ} with $I = A, B, C$	
AB	9.796
AC	4.887
BC	4.219

Table 12: Clayton parameter for three classes of obligors

7.2.2 Original Moments

The original moments are computed as in (6.1) and (6.2), where the copula $\mathbb{C}_{i;M}$ and the $\mathbb{C}_{i,j;M}$ are in the Clayton case rewritten as:

$$\mathbb{C}_{i;M} = p_i$$

$$\mathbb{C}_{i,j;M} = \left(p_i^{-\theta} + p_j^{-\theta} - 1\right)^{-\frac{1}{\theta}}.$$

This means that the first moment of the original distribution (6.1) is equivalent to the first moments of the original distribution in the Gaussian case (5.5). The second moments of the Gaussian case (5.6) differs from the second moments in the Clayton case, for the second terms in (6.2). The previous estimated parameters allow to compute the original moments, which are collected in **Table 13.**

Clayton: K-MEANS		
years	M_1	M_2
3	0.0070691	0.0036446
5	0.0207258	0.0106508
7	0.0403291	0.0210029
10	0.0733294	0.0382798

Table 13: Original Moments in the Clayton case

7.2.3 Moment Matching

Once the original moments have been computed, they are calibrated on the moments of the approximating distributions, by the MM procedure.

That is, by equating the original moments in (6.1) and (6.2) to the moments of the proxy distributions recovered in (6.8) and (6.9), the uniform survival probability q

and the uniform parameter θ are then obtained. **Table 14** collected the uniform parameters (q, θ) for different maturities.

Clayton: calibrated parameters			
years	q	$p = 1 - q$	θ
3	0.9929	0.0071	82.011
5	0.9793	0.0207	44.446
7	0.9597	0.0403	22.568
10	0.9267	0.0733	11.672

Table 14: Calibration result for the Clayton case

7.2.4 Pricing CDX Tranches

The final CDX tranches are computed by applying the spread formula in (2.7).

However, the ETL formula for the Clayton setup differs from the one used in the Gaussian framework. According to the preposition 6.3, I derived in the previous sections, the ETL for different maturities are collected in **Table 15.**

Clayton: ETL K-MEANS						
years	0-3%	3-7%	7-10%	10-15%	15-30%	30-100%
3	0.0414	0.0271	0.0157	0.0166	1.06E-02	2.29E-03
5	0.0883	0.0632	0.0387	0.0434	0.0313	9.97E-03
7	0.1664	0.1211	0.0746	0.0843	0.0611	1.98E-02
10	0.2931	0.2185	0.1358	0.1541	0.1124	3.62E-02

Table 15: ETL for the Clayton case

Finally, by using the ETL in equation (2.7) the tranches price spreads are obtained in **Table 16**.

Clayton: ETL K-MEANS					
0-3%	3-7%	7-10%	10-15%	15-30%	30-100%
5%	0.0228%	0.01356%	0.015497%	0.011049%	0.003419%

Table 16: Tranches spread for Clayton case

As usual, the spread of the equity tranche is fixed to 500 bps and the upfront charged on the equity tranche is equal to $U_{0-3}^{Clayton} = 0.20446$.

78

7.3   Results comments

In **Figure 14** the *K-means* tranches spread obtained in the Gaussian case and in the Clayton case are plotted. By comparing the tranches prices, it is possible to notice, that the Clayton approach leads to smaller equity and mezzanine tranches. The equity tranches, being the running spread fixed to 500 bps, is compared in terms of upfront. Then, the senior and the super senior tranches are bigger, when the dependence is modeled by Clayton copula. The comparison is more clear looking at the lower panel of the figure, where the y-axis are represented in log-scale.

The results obtained for the Clayton approach are only in parts coherent, with the *lower tail dependence*. As already reported in section 4.3 the tail dependence is strictly related with the marginal distributions. Roughly speaking, the tail dependence for a bivariate situation refers to the probability that a margins exceeds a certain thresholds, given that the other margin has already exceeded the same thresholds. In our framework, the margins distributions are the probability that the asset value of a firm is smaller than a certain thresholds. That is, when the asset value of a firm is smaller than the thresholds, there is a tendency of other asset value firms' to be smaller than their thresholds as well.

Generally speaking, this implies the tendency of defaults occurring in a chain situation.

According to this propriety the Clayton premia are expected to be greater than the Gaussian ones, because the greater is the probability of chain defaults, and so the greater is the expected loss, the greater is the amount requested for giving protection. However this evidence is only respected for the senior and super senior tranches, because in these tranches the *lower tail dependence* has more weight. In fact to reach such percentages of losses, the chain default, proper of the Clayton dependence, is fundamental.

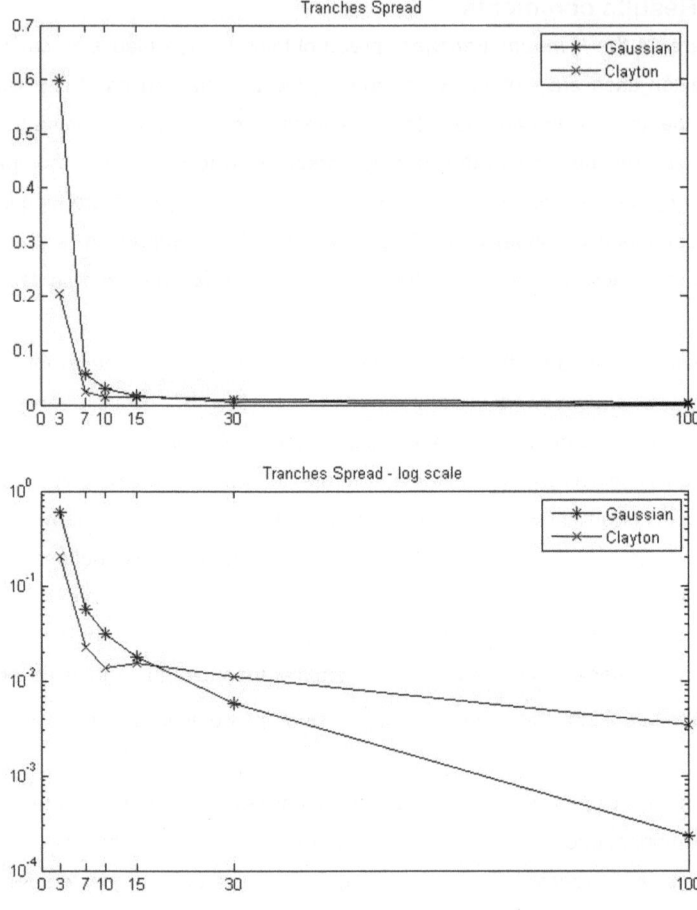

Figure 14: Tranches spreads for Gaussian and Clayton approach in linear and log-scale [34]

---

8 Conclusion

The MM method provides a computational methodology, which is based on sophisticated framework, but at the same time avoids huge numerical procedure, required by most complex models. It provides a closed formula structure, which makes the computation quick and easier.

Furthermore, the method is consistent with the analytical Credit VaR computation technique, presented by Castagna *et al.* (2009). This allows a consistent methodology to price the risk of CDO and to estimate the Credit VaR.

The difficulties of this technique lies in the original moments computation. In fact the number of inputs required to compute them is huge and a lot of assumptions are necessary for some of them.

Once the contagion terms have been eliminated and the asset dynamics dependencies are reduced by to one common factor, the implementation has been possible.

In addition, the large number of obligors in the portfolio increases the computation difficulty. Working with huge numbers of obligors is difficult both to collect all the time series required and to manage them in calculations.

A method to avoid these difficulties, proposed in the thesis, is clustering. However, grouping the obligors in classes and assuming homogenous classes, further increase the errors due to simplification.

The Clayton approach shares with the Gaussian approach the same problems of numerousness. The results obtained in the two methods, are consistent with the different dependence characteristic of the copula used, only for the higher tranches, such as the senior and the super senior.

9 Bibliography

Beinstein, E. / Scott, A. (2006): Credit Derivatives Handbook, Corporate
    Quantitative Research, JPMorgan.

Bluhm, C. (2003): CDO Modeling: Techniques, Examples and Application, Group
    Credit Portfolio Management, GCP3 - Structured Finance Analytics,
    HypoVereinsbank.

Bluhm, C. / Overbeck, L. / Wagner, C. (2003): An Introduction to Credit Risk
    Modeling, Chapman & Hall/CRC, Financial Mathematics Series; 2nd reprint.

Bluhm, C. / Overbeck, L. (2007): Structured Credit Portfolio Analysis Baskets &
    CDOs, Chapman & Hall/CRC, Financial Mathematics Series.

Castagna, A. / Mercurio, F. / Mosconi, P. (2009): Analytical credit var with
stochastic probabilities of default and recoveries.

Castagna, A. / Mercurio, F. / Mosconi, P. (2012): Analytical Pricing of CDOs in a
    Multi-Factor Setting by a Moment Matching Approach.

Cherubini, U. / Della Lunga, G. (2007): Structured Finance. The Object-Oriented
    Approach, John Wiley & Sons.

Cherubini, U. / Gobbi, F. / Mulinacci, S. / Romagnoli, S. (2011): Dynamic Copula
    Methods in Finance, The Wiley Finance Series, John Wiley and Sons.

Cherubini, U. / Luciano, E. / Vecchiato, W. (2004): Copula methods in finance,
    Wiley Finance Series, John Wiley & Sons.

Flanagan, C. / Sam, T. (2002): CDO Handbook, Global Structured Finance
    Research, JPMorgan.

Frees, E.W. / Valdez, E. (1998): Understanding relationship using copulas. North
    American Actuarial Journal, 2, 1-25.

Gordy, M.B. (2003): A risk-factor model foundation for ratings-based bank capital
    rules, Journal of Financial Intermediation, 12, 199–232.

Levy, E. (1992): Pricing european average rate currency options, Journal of
    International Money and Finance, 14, 474–491.

Marshall, A.W. / Olkin, I. (1988): Families of multivariate distribution, Journal of the
    American Statistical Association, 83, 834 -841.

McNeil, A. J. / Rüdiger, F. / Embrechts, P. (2006): Quantitative Risk
    Management: Concepts, Techniques and Tools, Princeton Univer. Press.

Merrill Lynch (2006): Credit Derivatives Handbook, Vol. 1-2.

Merton, R. (1974): On the pricing of corporate debt: The risk structure of interest rates. Journal of Finance, 2, 449–471.

Nelsen, R. B (1999): An Introduction to Copulas, Springer Series in Statistics.

Schlösser, A. (2010): Pricing and Risk Management of Synthetic CDOs, Springer.

Schönbucher, P. J. (2002): Taken to the Limit: Simple and not-so-simple Loa Loss Distribution, Working paper, Bonn University.

Schönbucher, P. J. (2003): Credit Derivatives Pricing Models, John Wiley & Sons.

Turnbull, S / Wakemann, L. (1992): A quick algorithm for pricing european average options, Journal of Financial and Quantitative Analysis, 26, 377–389.

Vasicek, O. (1991): Limiting loan loss probability distribution, KMW Corporation.

Wikipedia: The free encyclopedia. (2012, May 5). FL: Wikimedia Foundation, Inc. Retrieved May 10, 2004, from http://en.wikipedia.org/wiki/Self-organizing_map.

10 Appendix A

Composition of the reference portfolio and quotes of the CDS contracts in 3rd July 2007

CDS	1	2	3	4	5	6	7	8	9	10
ACE US	10.30	14.10	18.00	24.30	29.50	34.24	39.00	41.32	43.67	46.00
AET US	6.83	8.04	10.29	16.00	22.00	25.60	29.22	32.08	34.96	37.83
AL CN	9.70	13.40	17.00	23.11	31.90	37.43	43.00	48.51	54.06	59.60
AA US	14.06	18.45	24.09	34.33	43.70	52.08	60.50	67.50	74.56	81.60
AT US	80.67	115.99	195.13	237.50	329.00	364.40	400.00	414.93	429.99	445.00
MO US	13.30	15.82	19.50	21.78	28.00	34.13	40.30	45.51	50.76	56.00
AEP US	8.02	9.62	12.47	16.70	21.80	25.91	30.04	32.63	35.23	37.83
AXP US	7.80	10.14	13.90	16.40	18.00	20.04	22.10	24.42	26.77	29.10
AIG US	12.80	13.40	14.00	14.60	14.70	16.59	18.50	20.16	21.83	23.50
AMGN US	8.13	9.83	12.68	16.50	21.00	26.49	32.00	36.65	41.33	46.00
APC US	21.07	23.92	21.00	28.67	35.80	44.23	52.70	57.78	62.90	68.00
ARW US	14.60	19.40	25.23	36.70	47.80	58.87	70.00	77.96	85.99	94.00
T US	8.34	10.04	12.99	18.69	23.00	28.86	34.75	34.75	34.75	34.75
24004Z US	6.72	7.72	9.98	12.10	14.80	19.18	23.58	27.19	30.83	34.46
AZO US	8.99	11.41	18.00	23.63	33.90	42.18	50.50	57.47	64.49	71.50
BAX US	5.75	6.36	8.11	10.90	14.30	17.89	21.50	23.82	26.17	28.50
8891Z US	2.25	4.04	6.35	7.10	8.30	10.15	12.00	13.06	14.13	15.20
BSX US	24.95	34.44	45.03	60.50	77.00	89.90	102.87	111.25	119.70	128.12
BMY US	6.18	6.88	8.84	12.60	16.50	20.64	24.81	27.86	30.93	34.00
BNI US	7.70	10.30	12.90	18.38	25.50	32.73	40.00	44.31	48.66	53.00
CPB US	8.99	11.09	14.00	20.44	20.70	25.59	30.50	32.99	35.50	38.00
8125Z US	11.36	15.30	20.00	23.00	25.00	28.49	32.00	34.22	36.47	38.70
CAH US	10.93	13.83	17.97	26.00	34.50	43.73	53.00	59.30	65.66	72.00
CCL US	8.88	12.67	15.27	20.75	25.40	30.69	36.00	39.01	42.04	45.07
CAT US	9.53	9.30	10.29	13.90	18.20	22.69	27.20	30.52	33.86	37.20
CBS US	16.43	22.03	28.65	38.75	53.70	65.61	77.58	77.58	77.58	77.58
CTX US	41.86	62.59	83.20	109.63	129.20	141.62	154.10	161.86	169.69	177.50
CTL US	16.11	21.50	28.03	39.27	53.00	64.09	75.23	84.76	94.37	103.95
CI US	6.94	8.04	10.40	16.60	23.00	26.92	30.86	34.09	37.34	40.58
CIT US	19.99	26.97	35.00	40.09	47.00	50.74	54.50	57.75	61.03	64.30
15659Z US	8.99	11.09	14.34	20.03	26.00	32.97	39.98	45.02	50.11	55.18
CSC US	25.81	35.70	46.69	67.70	89.70	104.16	118.70	132.83	147.09	161.30
CAG US	9.20	12.77	15.50	21.16	28.00	34.98	42.00	46.31	50.66	55.00
COP US	7.15	9.41	12.78	15.60	20.00	22.99	26.00	28.85	31.73	34.60
CEG US	9.42	11.62	15.07	19.31	26.70	33.37	40.08	44.87	49.69	54.50
8191Z US	35.07	46.31	56.40	65.71	74.20	78.59	83.00	86.25	89.53	92.80
COX US	8.99	10.98	14.24	19.00	25.60	32.97	40.39	44.75	49.15	53.54
CSX US	10.18	12.88	22.00	26.30	49.00	58.47	68.00	74.64	81.33	88.00
CVS US	9.31	11.51	14.86	20.95	28.80	36.38	44.00	49.97	56.00	62.00
DE US	5.64	6.57	10.92	15.09	18.80	23.12	27.47	30.16	32.87	35.58
DVN US	7.00	9.30	11.50	17.30	22.70	27.59	32.50	35.42	38.36	41.30

D US	9.64	10.04	14.03	17.25	21.10	25.04	29.00	32.48	36.00	39.50
DUK US	7.80	9.41	12.06	15.10	19.00	23.33	27.68	29.46	31.24	33.03
DD US	7.30	9.60	12.00	14.78	18.40	23.83	29.30	31.85	34.43	37.00
EMN US	14.17	18.56	24.19	32.27	43.90	53.92	64.00	70.74	77.53	84.30
EQ US	21.28	28.97	37.88	53.98	71.30	87.76	104.30	118.26	132.33	146.37
EXPE US	46.38	80.54	127.21	174.40	235.60	267.36	299.30	316.96	334.76	352.52
FRE US	3.90	6.46	8.63	10.35	9.40	10.79	12.19	13.24	14.31	15.37
FNM US	3.45	6.67	9.36	9.10	10.50	12.52	14.55	15.43	16.32	17.21
M US	17.19	22.66	36.00	55.01	74.30	89.11	104.00	114.62	125.32	136.00
FDC US	124.95	201.90	287.55	364.83	422.70	456.20	489.89	507.17	524.60	541.98
FE US	10.71	13.72	17.76	21.10	25.30	30.88	36.50	39.49	42.50	45.50
GELK US	8.88	11.19	12.30	15.81	17.70	19.20	20.70	21.80	22.90	24.00
GIS US	7.80	9.30	14.00	17.04	21.10	26.14	31.20	35.12	39.06	43.00
GR US	3.90	7.62	11.43	15.70	23.30	29.48	35.70	39.95	44.23	48.50
HAL US	6.20	8.20	10.30	15.50	20.70	25.09	29.50	32.98	36.50	40.00
HPQ US	5.97	6.57	8.43	12.50	17.30	20.17	23.06	23.06	23.06	23.06
HON US	7.50	10.20	13.00	16.70	19.90	25.93	32.00	35.95	39.93	43.90
IACI US	35.93	50.41	66.06	95.60	126.20	148.22	170.36	183.58	196.91	210.20
	19.77	22.56	26.58	39.60	53.50	66.12	78.80	85.83	92.93	100.00
IBM US	6.40	7.30	9.36	13.00	17.30	20.07	22.86	24.44	26.03	27.62
ILFC US	8.23	9.41	13.80	17.50	20.70	22.70	24.70	26.69	28.70	30.70
IP US	9.42	15.90	23.00	32.79	39.80	50.12	60.50	66.64	72.83	79.00
JCP US	14.06	18.56	29.50	39.89	54.50	67.22	80.00	87.96	95.99	104.00
JNY US	29.26	39.27	66.00	92.04	118.00	137.70	157.50	170.61	183.82	197.00
KFT US	11.69	14.98	20.00	25.79	36.30	45.33	54.40	61.14	67.93	74.70
LEN US	30.33	48.31	77.50	102.53	120.60	131.27	142.00	149.70	157.46	165.20
LTD US	14.92	19.82	32.50	47.19	60.50	72.07	83.70	93.49	103.36	113.20
LMT US	3.45	5.30	8.00	12.20	16.30	20.54	24.80	27.09	29.40	31.70
LTR US	7.30	9.60	12.00	14.50	15.50	19.33	23.17	25.44	27.72	30.00
MAR US	11.25	14.67	19.42	31.35	41.80	53.56	65.39	72.89	80.46	88.00
MMC US	26.20	36.90	47.50	60.80	71.70	87.31	103.00	112.95	122.99	133.00
16302Z US	17.40	26.03	35.00	39.37	51.60	53.79	56.00	57.89	59.80	61.70
MCD US	8.13	8.56	10.00	16.22	17.00	22.09	27.20	30.45	33.73	37.00
MCK US	12.22	15.93	20.67	29.00	37.80	47.97	58.20	64.70	71.26	77.80
MWV US	13.30	24.10	34.70	52.13	63.00	75.72	88.50	98.72	109.02	119.30
MET US	5.70	7.90	10.00	12.90	15.30	17.89	20.50	22.49	24.50	26.50
MOT US	12.33	16.03	20.98	27.64	37.00	46.35	55.76	63.35	71.01	78.65
2381A US	5.86	6.46	8.32	10.20	13.10	16.43	19.78	21.51	23.26	25.00
NWL US	8.45	9.62	14.50	20.13	25.60	32.03	38.50	42.32	46.16	50.00
14408Z US	8.23	10.04	12.89	18.10	24.00	30.18	36.39	40.69	45.03	49.36
JWN US	8.13	9.83	14.00	16.84	22.40	29.18	36.00	39.98	44.00	48.00
NSC US	9.90	13.70	17.40	22.91	31.70	39.48	47.30	52.41	57.56	62.70
NOC US	5.50	7.00	8.50	12.90	16.70	20.69	24.70	27.06	29.43	31.80
OLN US	35.20	50.40	65.50	87.52	117.70	139.74	161.90	175.70	189.62	203.50
OMC US	6.62	7.62	9.77	14.20	18.80	24.10	29.42	33.63	37.88	42.12
PGN US	7.26	8.56	11.02	14.10	17.80	21.65	25.52	27.95	30.39	32.83
PHM US	45.30	68.79	105.00	140.48	172.50	185.71	199.00	209.39	219.86	230.30

RRD US	24.62	34.02	44.41	68.59	81.20	95.63	110.14	120.91	131.77	142.60
RDN US	28.20	39.90	51.50	65.80	77.70	81.89	86.10	89.09	92.10	95.10
RTN US	7.05	8.25	10.60	13.60	16.80	20.79	24.80	27.19	29.60	32.00
578903Z US	169.90	194.70	203.50	216.30	229.30	238.33	247.40	253.44	259.53	265.60
ROH US	9.00	12.30	15.50	20.13	25.80	32.38	39.00	44.64	50.33	56.00
SWY US	10.71	13.19	19.53	27.44	39.10	49.02	59.00	65.64	72.33	79.00
SLE US	10.50	14.77	23.50	31.86	42.40	52.44	62.53	70.32	78.17	86.00
SRE US	8.77	10.77	13.92	17.50	22.00	27.08	32.19	34.97	37.78	40.58
12968Z US	10.39	13.09	16.93	19.82	26.00	28.73	31.47	34.56	37.68	40.79
LUV US	10.82	14.14	22.22	35.46	50.00	61.57	73.20	80.93	88.73	96.50
S US	20.42	27.71	36.22	50.28	61.70	72.83	84.03	94.02	104.10	114.15
HOT US	26.02	42.63	73.31	108.29	141.00	170.31	199.78	215.44	231.24	246.99
TGT US	6.83	6.88	11.95	15.60	20.50	25.87	31.27	34.94	38.63	42.32
TIN US	18.80	35.90	51.70	73.43	92.40	107.81	123.30	134.78	146.36	157.90
3339Z US	6.40	7.30	9.36	12.30	16.00	18.74	21.50	24.15	26.83	29.50
ALL US	6.80	9.10	11.50	14.30	16.10	19.54	23.00	24.33	25.67	27.00
CB US	7.50	9.70	12.00	14.30	16.00	18.49	21.00	22.99	25.00	27.00
DOW US	6.62	10.90	15.50	18.89	26.40	32.23	38.10	44.17	50.29	56.40
HIG US	6.30	8.60	11.00	13.50	16.00	18.24	20.50	22.66	24.83	27.00
KR US	10.50	14.35	22.00	31.45	42.80	53.61	64.47	72.78	81.15	89.50
SHW US	12.70	17.40	22.00	27.95	34.30	40.13	46.00	50.31	54.66	59.00
DIS US	6.08	6.78	8.74	12.20	16.30	21.57	26.86	29.79	32.74	35.69
TWX US	9.64	12.04	15.58	22.39	26.80	34.55	42.34	48.16	54.02	59.87
TOL US	26.56	44.00	82.30	105.83	126.80	139.62	152.50	161.89	171.36	180.80
RIG US	11.04	11.70	13.00	19.10	24.30	31.08	37.90	42.38	46.90	51.40
TSN US	18.05	24.66	35.70	51.82	66.80	82.85	98.98	109.29	119.69	130.06
UNP US	15.90	17.20	18.50	19.00	33.00	41.23	49.50	54.98	60.50	66.00
UHS US	23.01	31.70	41.41	59.30	78.50	93.56	108.70	118.33	128.03	137.71
VLO US	9.60	13.40	17.20	24.14	31.80	38.33	44.90	48.95	53.03	57.10
VZ US	7.37	8.78	11.23	16.53	22.00	27.49	33.01	37.72	42.48	47.22
WMT US	5.97	5.09	8.11	9.01	11.00	13.08	15.17	16.63	18.10	19.56
WM US	14.38	22.20	31.50	38.60	45.30	48.24	51.20	54.45	57.73	61.00
WFC US	8.88	8.50	9.30	11.70	13.50	14.75	16.00	17.59	19.20	20.80
WY US	10.61	19.00	28.00	40.81	51.50	64.37	77.30	86.82	96.43	106.00
WHR US	9.85	12.35	17.97	32.58	38.80	49.66	60.58	67.46	74.39	81.30
WYE US	6.18	7.09	9.05	10.90	13.50	16.22	18.96	21.30	23.65	26.00
XL US	12.30	17.10	22.00	30.50	38.40	42.69	47.00	49.99	53.00	56.00

11   Appendix B

Bootstrapped defaults probabilities

CDS	1	2	3	4	5	6	7	8	9	10
ACE US	0.002	0.005	0.009	0.016	0.025	0.035	0.046	0.056	0.067	0.078
AET US	0.001	0.003	0.005	0.011	0.019	0.026	0.035	0.044	0.054	0.065
AL CN	0.002	0.004	0.009	0.016	0.027	0.038	0.051	0.066	0.083	0.102
AA US	0.002	0.006	0.012	0.023	0.037	0.053	0.072	0.091	0.114	0.138
AT US	0.013	0.038	0.096	0.153	0.260	0.334	0.413	0.470	0.528	0.584
MO US	0.002	0.005	0.010	0.015	0.024	0.035	0.048	0.062	0.078	0.097
AEP US	0.001	0.003	0.006	0.011	0.018	0.026	0.036	0.045	0.054	0.065
AXP US	0.001	0.003	0.007	0.011	0.015	0.020	0.026	0.033	0.041	0.050
AIG US	0.002	0.004	0.007	0.010	0.012	0.017	0.022	0.027	0.034	0.040
AMGN US	0.001	0.003	0.006	0.011	0.018	0.027	0.038	0.051	0.065	0.080
APC US	0.004	0.008	0.010	0.019	0.030	0.045	0.063	0.078	0.096	0.116
ARW US	0.002	0.006	0.013	0.025	0.040	0.060	0.083	0.105	0.131	0.158
T US	0.001	0.003	0.007	0.013	0.019	0.029	0.042	0.047	0.053	0.058
24004Z US	0.001	0.003	0.005	0.008	0.012	0.020	0.028	0.038	0.048	0.061
AZO US	0.001	0.004	0.009	0.016	0.029	0.043	0.060	0.079	0.100	0.123
BAX US	0.001	0.002	0.004	0.007	0.012	0.018	0.026	0.033	0.041	0.050
8891Z US	0.000	0.001	0.003	0.005	0.007	0.010	0.014	0.018	0.022	0.026
BSX US	0.004	0.011	0.022	0.040	0.064	0.090	0.120	0.147	0.177	0.209
BMY US	0.001	0.002	0.004	0.008	0.014	0.021	0.030	0.039	0.048	0.059
BNI US	0.001	0.003	0.006	0.012	0.022	0.034	0.048	0.061	0.075	0.091
CPB US	0.001	0.004	0.007	0.014	0.017	0.026	0.036	0.045	0.055	0.065
8125Z US	0.002	0.005	0.010	0.015	0.021	0.029	0.038	0.046	0.056	0.066
CAH US	0.002	0.005	0.009	0.017	0.029	0.045	0.063	0.081	0.101	0.123
CCL US	0.001	0.004	0.008	0.014	0.021	0.031	0.043	0.053	0.065	0.077
CAT US	0.002	0.003	0.005	0.009	0.015	0.023	0.033	0.042	0.053	0.065
CBS US	0.003	0.007	0.014	0.026	0.045	0.066	0.092	0.103	0.115	0.126
CTX US	0.007	0.021	0.041	0.072	0.106	0.138	0.173	0.205	0.239	0.275
CTL US	0.003	0.007	0.014	0.026	0.045	0.065	0.089	0.114	0.143	0.175
CI US	0.001	0.003	0.005	0.011	0.020	0.028	0.037	0.047	0.058	0.070
CIT US	0.003	0.009	0.017	0.027	0.039	0.051	0.063	0.077	0.091	0.107
15659Z US	0.001	0.004	0.007	0.013	0.022	0.034	0.048	0.062	0.078	0.095
CSC US	0.004	0.012	0.023	0.045	0.075	0.104	0.137	0.175	0.216	0.262
CAG US	0.002	0.004	0.008	0.014	0.024	0.036	0.050	0.063	0.078	0.094
COP US	0.001	0.003	0.006	0.010	0.017	0.023	0.031	0.039	0.049	0.060
CEG US	0.002	0.004	0.008	0.013	0.023	0.034	0.048	0.062	0.077	0.094
8191Z US	0.006	0.015	0.028	0.043	0.061	0.077	0.095	0.112	0.131	0.150
COX US	0.001	0.004	0.007	0.013	0.022	0.034	0.049	0.062	0.076	0.092
CSX US	0.002	0.004	0.011	0.018	0.042	0.060	0.081	0.101	0.124	0.148
CVS US	0.002	0.004	0.007	0.014	0.024	0.037	0.053	0.069	0.087	0.107
DE US	0.001	0.002	0.005	0.010	0.016	0.024	0.033	0.041	0.051	0.062
DVN US	0.001	0.003	0.006	0.012	0.019	0.028	0.039	0.049	0.059	0.071
D US	0.002	0.003	0.007	0.012	0.018	0.025	0.035	0.045	0.056	0.068

DUK US	0.001	0.003	0.006	0.010	0.016	0.024	0.033	0.040	0.048	0.057
DD US	0.001	0.003	0.006	0.010	0.016	0.024	0.035	0.044	0.053	0.064
EMN US	0.002	0.006	0.012	0.022	0.037	0.055	0.076	0.096	0.118	0.143
EQ US	0.004	0.010	0.019	0.036	0.060	0.088	0.122	0.158	0.198	0.242
EXPE US	0.008	0.027	0.063	0.114	0.190	0.253	0.323	0.380	0.439	0.499
FRE US	0.001	0.002	0.004	0.007	0.008	0.011	0.014	0.018	0.022	0.027
FNM US	0.001	0.002	0.005	0.006	0.009	0.013	0.017	0.021	0.025	0.030
M US	0.003	0.008	0.018	0.037	0.062	0.090	0.122	0.152	0.186	0.223
FDC US	0.021	0.066	0.139	0.229	0.320	0.398	0.478	0.541	0.602	0.661
FE US	0.002	0.005	0.009	0.014	0.021	0.031	0.044	0.054	0.065	0.078
GELK US	0.001	0.004	0.006	0.011	0.015	0.019	0.024	0.029	0.035	0.041
GIS US	0.001	0.003	0.007	0.011	0.018	0.027	0.037	0.048	0.061	0.075
GR US	0.001	0.003	0.006	0.011	0.020	0.030	0.043	0.055	0.069	0.084
HAL US	0.001	0.003	0.005	0.010	0.018	0.026	0.035	0.045	0.057	0.069
HPQ US	0.001	0.002	0.004	0.008	0.015	0.021	0.028	0.031	0.035	0.039
HON US	0.001	0.003	0.007	0.011	0.017	0.027	0.039	0.050	0.062	0.076
IACI US	0.006	0.017	0.033	0.063	0.104	0.146	0.194	0.235	0.280	0.327
	0.003	0.008	0.013	0.027	0.045	0.067	0.093	0.115	0.140	0.167
IBM US	0.001	0.002	0.005	0.009	0.015	0.020	0.027	0.033	0.040	0.048
ILFC US	0.001	0.003	0.007	0.012	0.017	0.023	0.029	0.036	0.044	0.053
IP US	0.002	0.005	0.012	0.022	0.033	0.051	0.072	0.090	0.111	0.134
JCP US	0.002	0.006	0.015	0.027	0.046	0.068	0.095	0.118	0.145	0.174
JNY US	0.005	0.013	0.033	0.061	0.098	0.136	0.180	0.220	0.263	0.309
KFT US	0.002	0.005	0.010	0.017	0.031	0.046	0.065	0.083	0.104	0.128
LEN US	0.005	0.016	0.039	0.068	0.099	0.128	0.160	0.191	0.224	0.258
LTD US	0.002	0.007	0.016	0.032	0.051	0.073	0.098	0.125	0.156	0.189
LMT US	0.001	0.002	0.004	0.008	0.014	0.021	0.030	0.037	0.046	0.055
LTR US	0.001	0.003	0.006	0.010	0.013	0.020	0.028	0.035	0.043	0.052
MAR US	0.002	0.005	0.010	0.021	0.035	0.055	0.078	0.099	0.123	0.149
MMC US	0.004	0.012	0.024	0.040	0.060	0.087	0.120	0.150	0.182	0.218
16302Z US	0.003	0.009	0.017	0.026	0.043	0.054	0.065	0.077	0.089	0.101
MCD US	0.001	0.003	0.005	0.011	0.014	0.023	0.033	0.042	0.053	0.064
MCK US	0.002	0.005	0.010	0.019	0.032	0.049	0.069	0.088	0.109	0.132
MWV US	0.002	0.008	0.017	0.035	0.053	0.076	0.104	0.132	0.164	0.198
MET US	0.001	0.003	0.005	0.009	0.013	0.018	0.024	0.031	0.038	0.046
MOT US	0.002	0.005	0.010	0.019	0.031	0.047	0.066	0.086	0.109	0.135
2381A US	0.001	0.002	0.004	0.007	0.011	0.017	0.024	0.030	0.036	0.043
NWL US	0.001	0.003	0.007	0.014	0.022	0.033	0.046	0.058	0.071	0.086
14408Z US	0.001	0.003	0.006	0.012	0.020	0.031	0.044	0.056	0.070	0.085
JWN US	0.001	0.003	0.007	0.011	0.019	0.030	0.043	0.055	0.068	0.083
NSC US	0.002	0.005	0.009	0.015	0.027	0.040	0.057	0.072	0.089	0.107
NOC US	0.001	0.002	0.004	0.009	0.014	0.021	0.030	0.037	0.046	0.055
OLN US	0.006	0.017	0.033	0.058	0.097	0.138	0.185	0.226	0.271	0.319
OMC US	0.001	0.003	0.005	0.010	0.016	0.025	0.035	0.047	0.059	0.074
PGN US	0.001	0.003	0.006	0.009	0.015	0.022	0.031	0.038	0.047	0.057
PHM US	0.008	0.023	0.052	0.092	0.140	0.178	0.220	0.260	0.302	0.345
RRD US	0.004	0.011	0.022	0.046	0.068	0.095	0.128	0.159	0.194	0.232

RDN US	0.005	0.013	0.026	0.044	0.064	0.081	0.099	0.116	0.134	0.153
RTN US	0.001	0.003	0.005	0.009	0.014	0.021	0.030	0.037	0.046	0.055
578903Z US	0.028	0.064	0.098	0.137	0.179	0.219	0.260	0.299	0.337	0.376
ROH US	0.001	0.004	0.008	0.013	0.022	0.033	0.047	0.061	0.078	0.097
SWY US	0.002	0.004	0.010	0.018	0.033	0.050	0.070	0.089	0.111	0.134
SLE US	0.002	0.005	0.012	0.021	0.036	0.053	0.074	0.096	0.120	0.146
SRE US	0.001	0.004	0.007	0.012	0.019	0.028	0.039	0.048	0.058	0.070
12968Z US	0.002	0.004	0.008	0.013	0.022	0.029	0.037	0.047	0.058	0.070
LUV US	0.002	0.005	0.011	0.024	0.042	0.063	0.087	0.109	0.135	0.162
S US	0.003	0.009	0.018	0.034	0.052	0.073	0.098	0.126	0.156	0.190
HOT US	0.004	0.014	0.037	0.072	0.116	0.167	0.226	0.274	0.324	0.378
TGT US	0.001	0.002	0.006	0.010	0.017	0.026	0.038	0.048	0.060	0.073
TIN US	0.003	0.012	0.026	0.049	0.077	0.107	0.142	0.176	0.214	0.254
3339Z US	0.001	0.002	0.005	0.008	0.014	0.019	0.026	0.033	0.042	0.051
ALL US	0.001	0.003	0.006	0.010	0.014	0.020	0.028	0.033	0.040	0.046
CB US	0.001	0.003	0.006	0.010	0.013	0.019	0.025	0.031	0.039	0.047
DOW US	0.001	0.004	0.008	0.013	0.022	0.033	0.046	0.061	0.078	0.098
HIG US	0.001	0.003	0.006	0.009	0.013	0.019	0.024	0.031	0.038	0.047
KR US	0.002	0.005	0.011	0.021	0.036	0.055	0.077	0.099	0.124	0.152
SHW US	0.002	0.006	0.011	0.019	0.029	0.041	0.054	0.068	0.083	0.100
DIS US	0.001	0.002	0.004	0.008	0.014	0.022	0.033	0.041	0.051	0.062
TWX US	0.002	0.004	0.008	0.015	0.023	0.035	0.051	0.066	0.084	0.104
TOL US	0.004	0.015	0.041	0.070	0.104	0.136	0.172	0.206	0.243	0.281
RIG US	0.002	0.004	0.006	0.013	0.021	0.032	0.046	0.058	0.073	0.089
TSN US	0.003	0.008	0.018	0.035	0.056	0.083	0.116	0.146	0.179	0.215
UNP US	0.003	0.006	0.009	0.013	0.028	0.042	0.059	0.075	0.093	0.113
UHS US	0.004	0.011	0.021	0.040	0.066	0.094	0.127	0.156	0.189	0.224
VLO US	0.002	0.004	0.009	0.016	0.027	0.039	0.054	0.067	0.081	0.097
VZ US	0.001	0.003	0.006	0.011	0.019	0.028	0.040	0.052	0.066	0.082
WMT US	0.001	0.002	0.004	0.006	0.009	0.013	0.018	0.023	0.028	0.034
WM US	0.002	0.007	0.016	0.026	0.038	0.048	0.060	0.073	0.087	0.102
WFC US	0.001	0.003	0.005	0.008	0.011	0.015	0.019	0.024	0.030	0.036
WY US	0.002	0.006	0.014	0.027	0.043	0.065	0.092	0.117	0.146	0.178
WHR US	0.002	0.004	0.009	0.022	0.033	0.051	0.072	0.092	0.114	0.138
WYE US	0.001	0.002	0.005	0.007	0.011	0.017	0.023	0.029	0.037	0.045
XL US	0.002	0.006	0.011	0.020	0.032	0.043	0.055	0.067	0.080	0.094

12 Appendix C

Discount factors

Term	1	2	3	4	5	6	7	8	9	10
Rates	4.63%	4.57%	4.57%	4.59%	4.60%	4.700%	4.800%	4.900%	5.000%	5.000%
Discount	0.9557	0.9145	0.8745	0.8358	0.7986	0.7591	0.7202	0.6820	0.6446	0.6139